Iris Vela

CFD prediction of pool fires

Iris Vela

CFD prediction of pool fires

CFD prediction of thermal radiation of large, sooty, hydrocarbon pool fires

Südwestdeutscher Verlag für Hochschulschriften

Impressum/Imprint (nur für Deutschland/ only for Germany)

Bibliografische Information der Deutschen Nationalbibliothek: Die Deutsche Nationalbibliothek verzeichnet diese Publikation in der Deutschen Nationalbibliografie; detaillierte bibliografische Daten sind im Internet über http://dnb.d-nb.de abrufbar.
Alle in diesem Buch genannten Marken und Produktnamen unterliegen warenzeichen-, marken- oder patentrechtlichem Schutz bzw. sind Warenzeichen oder eingetragene Warenzeichen der jeweiligen Inhaber. Die Wiedergabe von Marken, Produktnamen, Gebrauchsnamen, Handelsnamen, Warenbezeichnungen u.s.w. in diesem Werk berechtigt auch ohne besondere Kennzeichnung nicht zu der Annahme, dass solche Namen im Sinne der Warenzeichen- und Markenschutzgesetzgebung als frei zu betrachten wären und daher von jedermann benutzt werden dürften.

Verlag: Südwestdeutscher Verlag für Hochschulschriften Aktiengesellschaft & Co. KG
Dudweiler Landstr. 99, 66123 Saarbrücken, Deutschland
Telefon +49 681 37 20 271-1, Telefax +49 681 37 20 271-0, Email: info@svh-verlag.de
Zugl.: Essen, Universität Duisburg-Essen, Diss. 2009

Herstellung in Deutschland:
Schaltungsdienst Lange o.H.G., Zehrensdorfer Str. 11, D-12277 Berlin
Books on Demand GmbH, Gutenbergring 53, D-22848 Norderstedt
Reha GmbH, Dudweiler Landstr. 99, D- 66123 Saarbrücken
ISBN: 978-3-8381-0917-6

Imprint (only for USA, GB)

Bibliographic information published by the Deutsche Nationalbibliothek: The Deutsche Nationalbibliothek lists this publication in the Deutsche Nationalbibliografie; detailed bibliographic data are available in the Internet at http://dnb.d-nb.de.
Any brand names and product names mentioned in this book are subject to trademark, brand or patent protection and are trademarks or registered trademarks of their respective holders. The use of brand names, product names, common names, trade names, product descriptions etc. even without
a particular marking in this works is in no way to be construed to mean that such names may be regarded as unrestricted in respect of trademark and brand protection legislation and could thus be used by anyone.

Publisher:
Südwestdeutscher Verlag für Hochschulschriften Aktiengesellschaft & Co. KG
Dudweiler Landstr. 99, 66123 Saarbrücken, Germany
Phone +49 681 37 20 271-1, Fax +49 681 37 20 271-0, Email: info@svh-verlag.de

Copyright © 2008 Südwestdeutscher Verlag für Hochschulschriften Aktiengesellschaft & Co. KG and licensors
All rights reserved. Saarbrücken 2008

Produced in USA and UK by:
Lightning Source Inc., 1246 Heil Quaker Blvd., La Vergne, TN 37086, USA
Lightning Source UK Ltd., Chapter House, Pitfield, Kiln Farm, Milton Keynes, MK11 3LW, GB
BookSurge, 7290 B. Investment Drive, North Charleston, SC 29418, USA
ISBN: 978-3-8381-0917-6

Content

Abstract

Nomenclature

1 Introduction..1

2 Some characteristics and modeling of pool fires......................3

2.1 Burning velocity and mass burning rate..6

2.2 Flame geometry..11

2.2.1 Flame length..11

2.2.2 Flame tilt..16

2.2.3 Flame drag..18

2.3 Flow velocities..18

2.4 Flame temperature..20

2.5 Organized structures..24

2.5.1 Reactive zone...24

2.5.2 Hot spots..25

2.5.3 Soot parcels...26

2.6 Thermal radiation models 28

2.6.1 Semi-empirical radiation models...28

2.6.1.1 Point source radiation model (PSM) 29

2.6.1.2 Solid flame radiation models (SFM, MSFM) 31

2.6.1.3 Two zone radiation models (TZM) 34

2.6.2 Organized structures radiation models (OSRAMO II, OSRAMO III) 35

2.6.3 Radiation model according to Fay...40

2.6.4 Radiation model according to Ray...41

2.7 Irradiance..41

2.8 Field models and integral models..44

2.9 CFD simulation..45

2.10 Wind influence..48

3 Experiments...51

3.1 Pools...51

3.2 Fuels...52

3.3 IR thermographic camera system 52

3.4 VIS camera system...53
3.5 Radiometer measurements...53
3.6 Wind measurements..55
4 Some important topics of CFD used in this work 57
4.1 The conservation equations in fire modeling................................57
4.1.1 Overall mass conservation..59
4.1.2 Species mass conservation..60
4.1.3 Momentum conservation..61
4.1.4 Energy conservation...62
4.2 Sub-models in fire modeling...63
4.2.1 Modeling of turbulence...65
4.2.1.1 k-ε and k-ω models..67
4.2.1.2 Large Eddy Simulation (LES) 68
4.2.1.3 Scale Adaptive Simulation (SAS) 74
4.2.2 Combustion models..76
4.2.2.1 Source terms of species on the basis of chemical reactions............76
4.2.2.2 Eddy dissipation...77
4.2.2.3 Flamelet model 78
4.2.3 Radiation models..82
4.2.3.1 Photometric sizes and radiation balance equation 82
4.2.3.2 Discrete Ordinate 86
4.2.3.3 Monte Carlo 87
4.2.4 Soot models...88
4.2.4.1 Magnussen 88
4.2.4.2 Lindstedt 89
4.2.4.3 Tesner 92
4.3 ANSYS CFX and FLUENT software..94
4.3.1 Discretisation methods and solution algorithms................................94
4.3.1.1 Finite volume method..94
4.3.1.2 Geometry and mesh generation...96
4.4 Procedure of CFD simulation...99
4.4.1 Geometry and meshing 100
4.4.1.1 Geometry and meshing for the fire in calm condition 100

4.4.1.2 Geometry and meshing for the fire under the wind influence 101
4.4.2 Initial and boundary conditions and time steps.....................................102
4.4.3 Determination and configuration of sub-models...................................105
4.4.3.1 Sub-models for turbulence...105
4.4.3.2 Sub-models for combustion...107
4.4.3.3 Sub-models for thermal radiation..107
4.4.3.4 Sub-models for soot..109
4.4.4 Modeling of absorption coefficient of the flame...111
5 Results and discussions...113
5.1 Instantaneous and time averaged flame temperatures...................................113
5.1.1 Thermograms...113
5.1.2 Histograms..116
5.1.3 Probability density function (pdf) 117
5.1.4 Temperature fields..119
5.1.5 Axial and radial profiles..128
5.2 Instantaneous and time averaged Surface Emissive Power (SEP) 133
5.2.1 Four-step discontinuity function of temperature dependent
   absorption coefficient..133
5.2.2 Thermograms...135
5.2.3 Histograms..137
5.2.4 Probability density function (pdf) 140
5.2.5 Determination of SEP by isosurface of flame temperature.......................141
5.2.6 Integration of incident radiation G for determination of SEP....................146
5.2.7 Determination of SEP by irradiance as a function of distance...................147
5.3 Instantaneous and time averaged irradiance..148
5.3.1 Virtual radiometers... 148
5.3.2 Prediction of irradiance...149
5.4 Wind influence 152
5.4.1 Flame height, flame tilt, flame drag..152
5.4.2 Wind influence on surface emissive power (SEP), irradiance (E),
   temperature and flow velocity..162
5.5 Validation of CFD results..171
6 Conclusions...175
References..177

Abstract

Computational Fluid Dynamics (CFD) simulations of large-scale JP-4 pool fires with pool diameters of d = 2 m, 8 m, 16 m, 20 m and 25 m in a calm condition, as well as with pool diameters of d = 2 m, 20 m and 25 m under cross-wind conditions with wind velocities in a range of 0.7 m/s < u_w < 16 m/s are performed. CFD prediction of emission temperatures T, surface emissive power (SEP) and irradiances E($\Delta y/d$) at relative distances $\Delta y/d$ in horizontal direction from the pool rim is carried out.

Also, for the theoretical understanding of large pool fires, the time dependent flame temperatures are of great interest.

CFD predicted vertical temperature profiles $\overline{T}_{CFD}(x/d)$ for different relative radial distances y/d = 0, y/d = 0.05 and y/d = 0.1 show that the absolute maximum flame temperatures $\overline{T}_{max,CFD}(d)$ are away (y/d = 0.05) from the flame axis and depend on d: 1300 K (d = 2 m), 1250 K (d = 8 m), 1230 K (d = 16 m), 1200 K (d = 25 m) which agree well with the measured temperatures $\overline{T}_{max,exp}(d)$. CFD predicted radial temperature profiles $\overline{T}_{CFD}(r)$ dependent on x/d are in agreement with measurements. For pool fire with d = 25 m, at x/d = 0.125 bimodal profiles $\overline{T}_{CFD}(r)$ are found, while for x/d = 0.25 unimodal temperature profiles $\overline{T}_{CFD}(r)$ exist.

The CFD simulation of the "derived" quantity SEP requires a definition of the flame surface. The present work presents three different ways to predict SEP_{CFD}. The first way is to determine the flame surface by using an isosurface at a constant temperature. The second way considers that the flame surface results from the integration of many parallel two-dimensional distributions of incident radiation G(x, y) along the z-axis perpendicular to the xy-plane. In the third way a virtual wide-angle radiometer is defined at the pool rim and the irradiance E($\Delta y/d$) as a function of $\Delta y/d$ is simulated. To simulate the SEP, more exactly, a temperature dependent effective absorption coefficient $\bar{\bar{æ}}_{eff}(T)$ of the dissipative structures (reaction zones, hot spots and soot parcels) and air as a four-step discontinuity function is developed.

CFD predicted $\overline{SEP}_{CFD}(d)$ values of JP-4 pool fires, obtained by the third way, are: 105 kW/m² (d = 2 m), 65 kW/m² (d = 8 m), 45 kW/m² (d = 16 m) and 35 kW/m² (d =

x   Abstract

25 m). The $\overline{SEP}_{CFD}$ value for d = 2 m under predicts the $\overline{SEP}_{exp}$ by a factor of 0.8 whereas a good agreement is found between $\overline{SEP}_{CFD}(d)$ and $\overline{SEP}_{exp}(d)$ for d = 8 m, 16 m and 25 m. Based on the first way the $\overline{SEP}_{CFD}$ values agree well with the measured $\overline{SEP}_{exp}$ values if the flame surface temperature of 1100 K is used for d = 2 m, 500 K for d = 8 m and 400 K for d = 16 m and 25 m.

Instantaneous h(T), h(SEP) and time averaged histograms $\overline{h(T)}$, $\overline{h(SEP)}$, lead to probability density functions of the emission surface temperatures (flame temperatures) $pdf(\overline{T})$ and the surface emissive power $pdf(\overline{SEP})$, determined by the second way. For example, from the predicted $pdf(\overline{T}_{CFD})$ and $pdf(\overline{SEP}_{CFD})$ for d = 16 m the temperature and SEP are in the intervals of 648 K < $\overline{T}_{CFD}$ < 1100 K and 10 kW/m^2 < $\overline{SEP}_{CFD}$ < 80 kW/m^2. The measured values are in the intervals 633 K < $\overline{T}_{exp}$ < 1200 K and 9 kW/m^2 < $\overline{SEP}_{exp}$ < 114 kW/m^2. The CFD predicted functions pdf(T), pdf(SEP) are consistent with the measured pdfs.

CFD predicted time averaged irradiances $\overline{E}_{CFD}(\Delta y/d, d)$ under predicts the measured $\overline{E}_{exp}(\Delta y/d)$ at the pool rim $\Delta y/d = 0$ for d = 2 m by a factor of 0.8 and over predicts $\overline{E}_{exp}(\Delta y/d)$ up to the factor of 1.6 at $\Delta y/d = 0.5$ whereas for d = 8 m, 16 m and 25 m the irradiances $\overline{E}_{CFD}(\Delta y/d)$ agree well with the measured $\overline{E}_{exp}(\Delta y/d)$. For example, $\overline{E}_{exp}(\Delta y/d, d)$ as a function of d at $\Delta y/d = 0.5$ the following values are found: 28 kW/m^2 (d = 2 m), 18 kW/m^2 (d = 8 m) and 5 kW/m^2 (d = 25 m).

The wind influence on large pool fire is a complex phenomenon. CFD simulations show that the wind influences the flame length, flame tilt, flame drag, the flame temperatures T, the SEP and the irradiances E. With increasing wind velocity u_w from 4.5 m/s to 10 m/s $\overline{SEP}_{CFD}$ and $\overline{E}_{CFD}(\Delta y/d)$ at the pool rim increase downwind by a factor of about 2 – 6 for d = 2 m and by a factor of about 2 – 7 for d = 20 m. In both cases $\overline{E}_{CFD}(\Delta y/d)$ do not increase if u_w > 10 m/s as it is found in experiments. In the upper section of the flames, depending on the flame tilt and drag, a decrease of flame temperature of several hundreds K is found.

With increasing wind velocity $u_w \geq 2.3$ m/s the predicted flame tilt from the vertical becomes significant. The CFD results show that two counter rotating vortices at the leeward side of the fire are formed at the minimum $u_w = 1.4$ m/s as observed in experiments. The predicted flame tilt and drag for $d = 20$ m begins from $20°$ and 1.1 at $u_w = 1.4$ m/s and ends with $80°$ and 2.5 at $u_w = 16$ m/s which agree with the experimental data. In a case of $d = 2$ m the flame tilt and drag reach values of $60°$ and 2.5 for $u_w = 4.5$ m/s and $80°$ and 2.8 for $u_w = 16$ m/s as in the experiments.

Flame temperatures T, surface emissive power SEP, irradiances E and the wind influence on large pool fires were at the first time predicted with CFD simulations. The CFD predictions are generally in good agreement with the measured values.

The CFD simulations allow (for future), the estimation of wind effects and also the important influence of multiple fires on the hazard potential.

The present work has, also shown that the hazard potential of large pool fires with CFD simulations of the thermal radiation can be estimated much better than before.

xii Abstract

Zusammenfassung

CFD-Simulationen (Computational Fluid Dynamics) großer JP-4-Poolfeuer mit Pooldurchmessern von d = 2 m, 8 m, 16 m, 20 m und 25 m wurden bei windstillen Bedingungen sowie mit Pooldurchmessern von d = 2 m, 20 m und 25 m bei Windeinfluss mit Windgeschwindigkeiten im Bereich von 0.7 m/s < u_w < 16 m/s durchgeführt. Die CFD-Vorhersagen von Emissionstemperaturen T, spezifischen Ausstrahlungen (SEP) und Bestrahlungsstärken E(Δy/d) abhängig von relativen Abständen Δy/d in horizontaler Richtung vom Poolrand stehen in guter Übereinstimmung mit Experimenten.

Für das theoretische Verständnis großer Poolbrände sind insbesondere auch die zeitabhängigen Flammentemperaturen von großem Interesse.

Die CFD vorhergesagten vertikalen Temperaturprofile $\overline{T}_{CFD}$(x/d) abhängig von den relativen radialen Abständen y/d = 0, 0.05 und 0.1 zeigen, dass die maximalen Flammentemperaturen $\overline{T}_{max,CFD}$(d) außerhalb (y/d = 0.05) der Flammenachse liegen und mit zunehmendem d wie folgt abnehmen: 1300 K (d = 2 m), 1250 K (d = 8 m), 1230 K (d = 16 m), 1200 K (d = 25 m). Die vorhersagten Temperaturen $\overline{T}_{max,CFD}$(d) sind in guter Übereinstimmung mit den gemessenen Temperaturen $\overline{T}_{max,exp}$(d). CFD vorhergesagte radiale Temperaturprofile $\overline{T}_{CFD}$(r) abhängig von x/d sind ebenfalls in guter Übereinstimmung mit Messungen. Für Poolfeuer (d = 25 m) bei x/d = 0.125 werden bimodale Profile $\overline{T}_{CFD}$(r) gefunden, während für x/d = 0.25 unimodale Temperaturprofile $\overline{T}_{CFD}$(r) existieren.

Die CFD-Berechnung der "abgeleiteten" Schlüsselgröße SEP benötigt eine Definition der Flammeoberfläche. In der vorliegenden Arbeit werden drei verschiedene Wege zur Vorhersage von SEP_{CFD} vorgestellt und diskutiert. Der erste Weg ist die Bestimmung der Flammenoberfläche, die als Isofläche konstanter Temperatur definiert ist. Die zweite Weg betrachtet die Flammenoberfläche als Ergebnis der Integration vieler paralleler zweidimensionaler Verteilungen G(x,y) der einfallenden Strahlung, entlang der z-Achse senkrecht zur xy-Ebene. Im dritten Weg wird ein virtuelles *Weitwinkel*-Radiometer am Poolrand positioniert und die Bestrahlungsstärke E(Δy/d) als Funktion der Abstands Δy/d berechnet. Zur genaueren

xiv Zusammenfassung

Simulation der SEP, wurde ein temperaturabhängiger effektiver Absorptionskoeffizient bezüglich dissipativer Strukturen (Reaktionszonen, hot spots und Rußballen) und heißer Luft, für eine vierstufige Funktion $\bar{\bar{æ}}_{eff}(T)$ entwickelt.

Die mit dem dritten Weg vorhergesagten $\overline{SEP}_{CFD}(d)$-Werte von JP-4-Poolfeuer sind: 105 kW/m² (d = 2 m), 65 kW/m² (d = 8 m), 45 kW/m² (d = 16 m) und 35 kW/m² (d = 25 m). Der $\overline{SEP}_{CFD}$-Wert für d = 2 m unterschätzt den experimentelle Wert um einen Faktor von 0.8, während eine gute Übereinstimmung zwischen $\overline{SEP}_{CFD}(d)$ und $\overline{SEP}_{exp}(d)$ für d = 8 m, 16 m und 25 m gefunden wurde. Basierend auf dem ersten Weg stimmen die $\overline{SEP}_{CFD}$-Werte gut mit den gemessenen $\overline{SEP}_{exp}$-Werten überein, wenn für die Flammenoberflächentemperaturen 1100 K (d = 2 m), 500 K (d = 8 m) und 400 K (d = 16 m und 25 m) gewählt werden.

Momentane h(T), h(SEP) und zeitlich gemittelte Histogramme $\overline{h}(T)$, $\overline{h}(SEP)$ führen zu Wahrscheinlichkeitsdichtefunktionen der Emissionstemperaturen (Flammentemperaturen) pdf($\overline{T}_{CFD}$) und der spezifischen Ausstrahlungen pdf($\overline{SEP}_{CFD}$), ermittelt mit dem zweiten Weg. Beispielsweise liegen die Temperaturen und SEPs aus den vorhergesagten pdf($\overline{T}_{CFD}$) und pdf($\overline{SEP}_{CFD}$) für d = 16 m in den Bereichen 648 K < $\overline{T}_{CFD}$ < 1100 K und 10 kW/m² < $\overline{SEP}_{CFD}$ < 80 kW/m². Die gemessenen Werte liegen in den Bereichen 633 K < $\overline{T}_{exp}$ < 1200 K und 9 kW/m² < $\overline{SEP}_{exp}$ < 114 kW/m². Die CFD-vorhergesagten Funktionen pdf(T), pdf(SEP) sind im Einklang mit den gemessenen pdfs.

CFD vorhergesagte, zeitlich gemittelte Bestrahlungsstärken $\overline{E}_{CFD}(\Delta y/d, d)$ unterschätzen die gemessenen $\overline{E}_{exp}(\Delta y/d)$ am Poolrand $\Delta y/d = 0$ für d = 2 m um einen Faktor von 0.8 und überschätzen $\overline{E}_{exp}(\Delta y/d)$ bis um den Faktor 1.6 bei $\Delta y/d$ = 0.5, während für d = 8 m, 16 m und 25 m die Bestrahlungsstärken $\overline{E}_{CFD}(\Delta y/d)$ gut mit den gemessenen $\overline{E}_{exp}(\Delta y/d)$ übereinstimmen. Beispielsweise sind für die

Funktion $\overline{E}_{exp}(\Delta y/d, d)$ bei $\Delta y/d = 0.5$ die folgenden Werte gemessen worden: 28 kW/m² (d = 2 m), 18 kW/m² (d = 8 m) und 5 kW/m² (d = 25 m).

Der Windeinfluss auf große Poolfeuer ist ein komplexes Phänomen. Die CFD-Simulationen zeigen, dass der Wind die Flammelänge, die Flammenneigung, das Flammendrag, die Flammentemperaturen, die SEP und die Bestrahlungsstärken E beinflusst. Mit zunehmender Windgeschwindigkeit von $u_w = 4.5$ m/s bis 10 m/s nehmen $\overline{SEP}_{CFD}$ und $\overline{E}_{CFD}(\Delta y/d)$ am Poolrand leewärts um einen Faktor von ca. 2 – 6 (für d = 2 m) oder um einen Faktor von ca. 2 – 7 (für d = 20 m) zu. In beiden Fällen nimmt $\overline{E}_{CFD}(\Delta y/d)$ nicht zu, wenn $u_w \geq 10$ m/s vorliegt, wie auch experimental gezeigt wurde. Im oberen Teil der Flamme wird abhängig von Flammenneigung und -drag, eine Abnahme der Flammentemperatur um mehrere 100 K gefunden.

Mit zunehmender Windgeschwindigkeit $u_w \geq 2.3$ m/s wird die vorhergesagte Flammenneigung signifikant. Die CFD-Ergebnisse zeigen, dass zwei gegenläufige Wirbel auf der Leeseite des Feuers gebildet werden bei einer minimalen Windgeschwindigkeit $u_w = 1.4$ m/s, wie auch experimentell beobachtet wurde. Die vorhergesagte Flammenneigung und das Flammedrag (für d = 20 m) beginnen ab 20° und 1.1 bei $u_w = 1.4$ m / s und enden bei 80° und 2.5 bei $u_w = 16$ m/s, in Übereinstimmung mit den experimentellen Daten. Für d = 2 m erreichen Flammenneigung und -drag Werte von 60° und 2.5 (bei $u_w = 4.5$ m/s) und 80° und 2.8 (bei $u_w = 16$ m/s), in Übereinstimmung mit den experimentellen Daten.

Flammentemperaturen T, spezifische Ausstrahlungen SEP, Bestrahlungsstärken E und Windeinflüsse auf große Poolfeuer wurden erstmals mit CFD-Simulationen vorhergesagt. Die CFD-Vorsagen stehen in der Regel in guter Übereinstimmung mit den gemessenen Werten.

Die CFD-Simulationen erlauben (in naher Zukunft), die Abschätzung des Windeinflusses sowie insbesondere den wichtigen Einfluss multipler Feuer auf das Gefährdungspotential.

Die vorliegende Arbeit hat auch gezeigt, dass das Gefährdungspotenzial großer Poolfeuer infolge der thermischen Strahlung mit CFD-Simulationen viel besser als bisher abgeschätzt werden kann.

Nomenclature

A	area (m^2)
A_F	flame area (m^2)
a_{hs}	area fraction of hot spots
a_x	area of a pixel, matrix element, in the thermographic image (m^2)
A_p	pool area (m^2)
a_{sp}	area fraction of soot parcels
A_T	isosurface of constant flame temperature (m^2)
c_p	specific heat capacity (kJ/(kg K))
d	pool diameter (m)
dA	infinitesimal area (m^2)
dG	infinitesimal incident radiation (kW/m^2)
ds	infinitesimal distance (m)
dV	infinitesimal volume (m^3)
E	irradiance (kW/m^2)
e	extinction coefficient
Fr	Froude number
Fr_f	flame Froude number
G	incident radiation (kW/m^2)
g	gravitational acceleration (m/s^2)
$g_T(T)$	pdf of temperature
$g_{SEP}(SEP)$	pdf of surface emissive power
h	specific enthalpy (kJ/kg)

xviii Nomenclature

H	length of the visible flame (m)
H_{cl}	length of the clear burning zone (m)
H/d	relative visible flame lenght
H_P	length of the fire plume (intermittency region of the flame (m))
H_{pul}	length of the pulsation flame zone (m)
h_{rim}	height of the pool rim (m)
$(-\Delta h_c)$	specific height of combustion (kJ/kg)
Δh_v	specific height of vaporization (kJ/kg)
I	radiation intensity (kW/(m^2 sr))
I_B	blackbody radiation intensity at temperature T (kW/(m^2 sr))
k	absorption coefficient (1/m)
l	length (m)
L	radiation intensity (W/(m^2 sr))
M	molar mass (kg/mol)
m_f	mass of fuel (kg)
$\dot{m}_f$	mass flow rate of the fuel (kg/s)
$\dot{m}_f''$	mass burning rate of the fuel (kg/(m^2 s))
N	particle number density in Lindstedt soot model (1/m^3)
N_s	number of species
N_T	number of total images in a thermographic sequence
n_0	spontaneous radical formation in Magnussen soot
p	pressure (bar)

p_a	ambient pressure (bar)
$\dot{Q}$	total heat release rate (kW)
$\dot{Q}_c$	heat of combustion (kW)
$\dot{Q}_b$	heat back from the flame to the pool surface (kW)
$\dot{q}$	thermal radiation per area (kW/m^2)
s	direction
SEP	surface emissive power (kW/m^2)
SEP_{act}	actual surface emissive power (kW/m^2)
SEP_{hs}	surface emissive power of hot spots (kW/m^2)
$SEP_{i,j}$	local emissive power of a pixel element (kW/m^2)
SEP_{LS}	surface emissive power of luminous spots (kW/m^2)
SEP_{sp}	surface emissive power of soot parcels (kW/m^2)
SEP_{SZ}	surface emissive power of soot zones (kW/m^2)
t	time (s)
Δt	time interval (s)
T	emission flame temperature (K)
T_a	ambient temperature (K)
$T_{i,j}$	temperature of the pixel element (K)
T_{in}	inlet temperature (K)
T_{max}	centerline maximum emission temperature (K)
t_b	burning time (s)

xx   Nomenclature

u	velocity (m/s)
v	velocity (m/s)
V	volume (m^3)
v_a	burning velocity of liquid fuel (m/s)
v_f	burning velocity of fuel (m/s)
X	specific concentration (mol/kg)
x	axial coordinate in vertical direction (m)
x_i, y_i	amount of component i
Y	mass fraction
y	radial coordinate in horizontal direction (m)
Δy	horizontal distance from the pool rim (m)
$\Delta y/d$	relative horizontal distance from the pool rim

Greek symbols

α	absorbance of the receiving area element
α	convective heat transfer (W/(m^2 K))
β	view angle (°)
β_F, β_E	view angles referring to a flame element, receiver element (°)
$\bar{\bar{æ}}_{eff}$	effective absorption coefficient of the flame (1/m)
ε	dissipation rate of turbulent kinetic energy (m^2/s^3)
ε_F	flame emissivity
φ	view factor (–)
λ	thermal conductivity coefficient (W/(m K))
ρ	density (kg/m^3)

ρ_a	density of air (kg/m^3)
ρ_f	density of fuel (kg/m^3)
ρ_s	density of soot (kg/m^3)
σ	Stephan Boltzman constant (5.67 × 10^{-8} W/(m^2 K^4))
τ	atmospheric transmittance (–)
Ω	solid angle (sr)

Indices

a	ambient
act	actual
B	black body
b	boiling
b	back
c	combustion
eff	effective
exp	experiment
F	flame
f	fuel
g	gas
hs	hot spots
i, j	the position of the pixel element in thermogram
LS	luminous spots
m	mass

xxii   Nomenclature

max	maximum
P	pool
P	plume
p	pressure
pul	pulsation
rad	radiation
rim	pool rim
s	soot
sp	soot parcels
SZ	soot zones
T	temperature
v	vaporization
w	wind

Miscellaneous

$(\overline{})$	time averaged value
$<>$	spatial averaged values
max(a,b)	maximum of a and b
i, j	the position of the pixel element in thermogram
OSRAMO	organized structures radiation model
pdf	probability density function
s	direction of propagation

1 Introduction

Accidental fire in process plants are often pool or tank fires e.g. Buncefield in December 2005 [1,2]. These fires show a potential risk for humans and neighboring facilities due to their thermal radiation, and formation of combustion products such as soot particles. Such fires are relatively little investigated experimentally [3-11] and especially with CFD simulations [9,12-19]. A key parameter for the prediction of thermal radiation of such fires is the Surface Emissive Power (SEP) [1,3-7,11,14-16,20-29]. It is usually defined as the heat flux due to the thermal radiation at the surface area of the flame. The SEP is dependent on the geometry, so for the prediction of SEP it is necessary to define a flame surface A_F [1,3-7,14-16,20-28]. In the past semi empirical models are used to determine the time average $\overline{SEP}$ value averaged over the whole flame surface $\overline{A}_F$ as it is assumed in the point source model (PSM) [5,7,25] and the solid flame model (SFM) [5,7,14-16,20-25]. The area $\overline{A}_F$ is usually assumed to be a cylinder or has a conical shape [5,7,14-16,20-25] or can be determined by pdf of different organized structures in a fire as is done in OSRAMO II, III [1,3,4,7,26-28]. The actual $\overline{SEP}_{act}$, the $\overline{SEP}_{hs}$ and the $\overline{SEP}_{sp}$ of hot spots and soot parcels can be obtained experimentally by evaluation of thermograms in combination with VIS images by detecting luminous and non-luminous zones [3,4,7,20,21,26-28]. The heat flux from the fire which can be received with radiometers at a certain distance from the flame is defined as irradiance $E(\Delta y/d)$ [3,4,7,16,25-28]. With the (empirical) radiation model according to Fay [22] and Raj [23,24] local profiles of thermal radiation can be considered. Ray [23,24] gives a criteria for the setting thermal radiation hazard zones around large hydrocarbon fires. He criticizes the current radiation models which do not consider the effect of the combustion dynamics associated with large size pool burning. According to Ray [24] the application of Computational Fluid Dynamics (CFD) to predict the dynamics and radiation from a realistic, large pool fire is still in its infancy.

CFD simulations of large pool fires are done, also to reduce the number of large-scale experiments. CFD simulation offers spatially and temporally resolution of thermal radiation inside and outside the fire as a function of the fire dynamics. University of Utah [16] and Sandia National Laboratory (SNL) [9,12,16] use CFD code to predict the heat flux to the container inside and outside the fire, Jensen et. al.

[29] validates different radiation models to predict heat flux profiles inside and outside the flame. By CFD the irradiances E(Δy/d) depending on relative horizontal distances Δy/d from the pool rim can be determined by virtual radiometers [3,4,7,27,28]. Some CFD simulations are done on large hydrocarbon pool fires (kerosene, JP-4) under the wind influence with different wind velocities [9,16,19]. Sinai et al. [19] investigate influence of the computational geometry on the predicted flame tilt and drag and a temperature as a consequence, SNL investigate and their influence of the flame tilt and drag on the temperature and thermal radiation from the fire to the surrounding, especially on the heat flux to the container involved in a fire [9,16,18].

In this work the CFD simulations of sooty, large, hydrocarbon pool fires e.g. JP-4 with d = 2 m, 8 m, 16 m, 20 m and 25 m are done to predict the emission temperatures (T), the surface emissive power (SEP) and the irradiances (E(Δy/d)). The large JP-4 pool fires with d = 2 m, 20 m and 25 m are also investigated by CFD under the influence of the cross wind with various wind velocities (0.7 m/s ≤ u_w ≤ 16 m/s) to predict the influence of flame tilt and drag on the $\overline{SEP}_{CFD}$ and $\overline{E}_{CFD}$ (Δy/d).

2 Some characteristics and modeling of pool fires

Pool fire is defined in the literature [25] as the combustion of material evaporating from a layer of liquid (fuel) at the base of the fire. It is generally turbulent, non-premixed, diffusion flame, which liquid fuel is spread out horizontally [7]. Pool fire is a kind of frequent accidental fires, which can occur in process industries by spontaneous release of liquid fuels during their storage, processing or transport. A fire in a liquid storage tank and a trench fire are also forms of a pool fire. A pool fire may also occur on the surface of flammable liquid spilled onto water. For the fire occurrence the relevant facts are the quantity of fuel in the fuel/air mixture, geometrical properties of the fire environment, temperature conditions and a heat transfer.

In the following chapter some physical and chemical properties of the pool fires are discussed.

There exists a different kinds of open fires produced by ignition of accidentally released flammable materials (liquids, droplets, gases or aerosols): pool fire, spill, tank fire, boilover fire, flare flame, jet flame, fire-gas/clouds, UVCE, BLEVE, fireball. The release scenarios, which can occur in chemical plants these types of fires have a significant hazard potential, particularly due to the heat radiation and convection, and the formation of combustion products (e.g. soot particles). The types of fires can be characterized as follows [7]:

a) *Pool, spill and tank fire*:

Pool fire is defined as the combustion of material, usually a liquid or a solid which can occur in a relatively thin layer on the surface of water or it can fill a pool.

In the case of fire spillage or leakage the flammable liquid spreads and form a spill on some surface (e.g. on the ground, plant area or on the water) without geometric limitations. In a case of tank fire a burning of the flammable substance usually occurs in the container such are individual tanks, tank farms, chemical reactors, columns or storage containers.

The pool fires, spills and the tank fires belong to non premixed flames (Fig. 2.1).

2 Some characteristics and modeling of pool fires

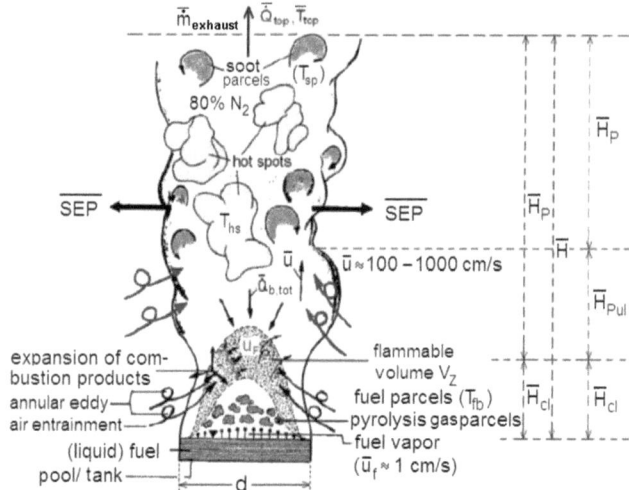

Fig. 2.1: Physical processes in adiabatic pool, spill and tank fires of liquid fuels [7].

b) *Boilover fire*:

It is an intense tank fire. The flammable liquid occurs on a layer of relatively low boiling liquids in a tank or tanks (e.g. oil on water layers in a storage tank). By spontaneous evaporation of low boiling liquid resulting from an overlying tank fire zone, large quantity of flammable liquid form a fireball from the tank or tank farm [7].

b) *Flare flames, jet flames*:

The combustible liquid or a combustible gas (mixed) occurs with a high momentum as the beam or jet into the atmosphere. In the case of a flare fire the release occur through a torch [7].

c) *Gas clouds fire, UVCE and BLEVE*:

They result from a leakage forming a combustible gas/air or steam/air mixture cloud, which in a certain time spreads and increases before it burns [7]. As a consequence can be either an atmospheric gas-clouds explosive (UVCE, Unconfined Vapor Cloud Explosion) or a burning gas clouds deflagrated fire or flash fire (UVCF Unconfined Vapor Cloud Fire) [7]. If there is an overheating (e.g. due to the heat radiation of a neighboring fire) of a tank or container which contains a pressurized, flammable liquid or a combustible, liquefied gas, a BLEVE (Boiling Liquid Expanding Vapor

Explosion) with typically intense pressure wave happen where e.g. tanks can break in individual parts (fragments) [7].

d) *Fireball*:

It happen by ignition of a flammable gas clouds of steam/air mixtures in the form of an unsteady, turbulent non-premixed flame, usually with a strong blast [7].

Pool fires, spills and tank fires occur as a 75% of the accidents in the process industries.

Accidents in the petrochemical industry occur due to e.g. spillage or leakage [28]. The heat release from a large flame effects as a thermal radiation on people and surrounding objects and can produce fatal injuries or damage the buildings or parts of the plant. In Fig. 2.2 is a tank fire in Buncefield accident in London 2005 [1-2]. The expansion of the fire to the neighbouring tanks happened without explosions due to the high thermal radiation.

Fig. 2.2: Buncefield fire

The Buncefield incident is a result of overfilling a very large mass of winter gasoline (m_f = 300 t in Buncefield), which led to a major fire of several days duration and involved 22 of a total of 41 tanks [1-2]. The analysis of the Buncefield incident so far has shown that the maximum visible relative flame height lies in the region of 2.5 < $(H/d)_{max,Bunc}$ < 6.5 and the predicted value lies in the region 1.8 < $(H/d)_{max,calc}$ < 1.9

[1]. For large, black smoky fires the estimation of the critical thermal separation distance is not dependent on the total fire, but on the height of a hot, clear burning zone. In addition, for multiple tank fires, there is a considerable increase in the mass burning rate, the flame height, the surface emissive power, as well as the thermal separation distance [1].

To contribute to the safe estimation of the heat radiation of the fire and to deal this work with other numerical calculations of the pool flames, a type of the open pool fire is presented.

(a)　　　　　　　(b)　　　　　　　(c)

Fig. 2.3: Types of open fires: (a) kerosene pool fire (d = 1.12 m) [31], (b) JP-4 pool fire (d = 16 m) [3,4], (c) n-pentane pool fire (d = 25) [3,4].

2.1 Burning velocity and mass burning rate

Calculation formula for time averaged burning velocity $\bar{v}_a$ Eq. (2.1a,b) exist from Hottel [32], Werthenbach [33], and Herzberg [34], whereby several assumptions and limitations must be made [35]. Calculations of $\bar{v}_{a,max}$ (Eq. 2.1b) have been made according to Burgess [36], also for several liquid mixtures according to Grumer [37]. For $v_a(d)$ by larger values of d large uncertainties exist. Calculation formula for $\bar{m}''_{f,max}$ exists according to Burgess [36], for $\bar{m}''_f(d)$ according to Zabetakis [38] as well as for $\bar{m}''_f(d)$ according to Babrauskas [39], which in each case contain empirical constants.

well as for $\overline{\dot{m}}_f''(d)$ according to Babrauskas [39], which in each case contain empirical constants.

One of the first systematic studies of the combustion behaviour of pool fires in dependence on fuel and pool diameter is done by Blinov Khudiakov [40], whose work have been later analysed by Hottel [32] who showed that as the pool (pan) diameter increases the fire regime changes from laminar to turbulent. Some results of this work are shown in Fig. 2.4 where the burning velocity v_a and a flame height h are plotted against the pool diameter.

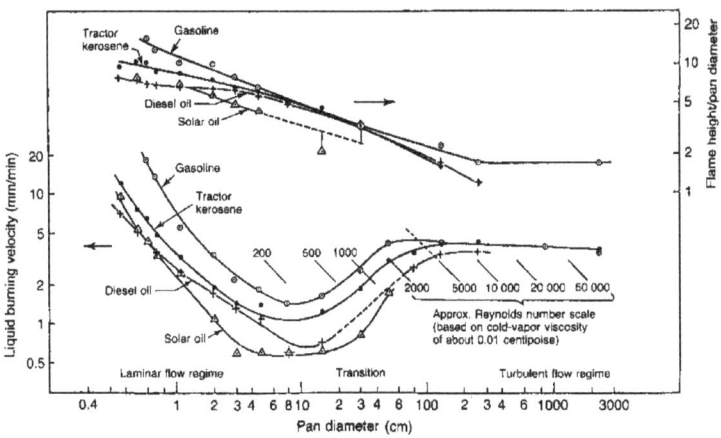

Fig. 2.4: Liquid burning velocity v_a and flame height as a function of fire regime depending on pool diameter d for various fuels according to Blinov Khudiakov [40] and Hottel [32].

v_a is a speed with which the fuel surface at a given volume of liquid drops, for all fuels tends to have the same dependence on the pool diameter d. Reynolds number Re is proportional to the product $v_a d$. In the laminar fire regime, for small start-up Reynolds numbers Re ≈ 20, with an increasing pool diameter d up to 0.1 m the burning rate v_a strongly decreases. In the transition fire regime for 20 < Re < 200, between laminar and turbulent, for 0.1 m ≤ d ≤ 1 m, the burning velocity v_a first increases, than decreases and finally levels off with increasing d but does not reach the maximum of the laminar regime. In the fully turbulent regime above Re = 500, at $d \geq 1$ m the burning velocity v_a remains constant with increasing d.

Hottel [32] found the connection between the fuel burning rate v_a, the heat feedback from the flame to the liquid pool $\overline{\dot{Q}}_{tot,b}$ and the combustion enthalpy Δh_V (Eq. (2.2a-d)):

$$\overline{v}_a\left(d, f, t, \Delta h_c / \Delta h_V, u_w\right) = \frac{\overline{\dot{Q}}_{tot,b}(d,f,t,(-\Delta h_c), u_w)}{A_p \rho_f \left(\overline{c}_{p,f}\left(T_{f,b} - T_{f,a}\right) + \Delta h_V\right)}. \qquad (2.1a)$$

with

$$\overline{v}_a(d) = \overline{v}_{a,max}(1 - e^{-\overline{æ}d}) \qquad \text{for } 0.4 \text{ cm} < d < 3000 \text{ cm} \qquad (2.1b)$$

$v_a = f(d, f, u_w, (-\Delta h_c)/\Delta h_V, t, \text{effects of pool rim})$ [7,35].

Crucial to the burning rate is the heat back from the flame to the pool surface. An energy balance on the pool surface helps in explaining this behaviour.

$$\overline{\dot{Q}}_{tot,b} = \overline{\dot{Q}}_{f,tot} \qquad (2.2a)$$

$$\overline{\dot{Q}}_{f,tot} = \overline{\dot{Q}}_V + \overline{\dot{Q}}_{\Delta T} + \overline{\dot{Q}}_{lost} \qquad (2.2b)$$

$$\overline{\dot{Q}}_{tot,b} = \overline{\dot{Q}}_{rad,b} + \overline{\dot{Q}}_{\alpha,b} + \overline{\dot{Q}}_{\lambda,b} \qquad (2.2c)$$

$$\overline{\dot{Q}}_{tot,b} = \frac{4\lambda}{d}(T_F - T_a) + \alpha(T_F - T_a) + \sigma(T_F^4 - T_a^4)(1 - e^{-\overline{æ}d}). \qquad (2.3)$$

The first term in Eq. (2.3) describes the heat flow through conduction along the tank wall, the second term, the convective heat back flow and the third, radiation transport to the liquid fuel. Accordingly, λ symbolize the thermal conductivity coefficient, α convective heat transfer coefficient, $\overline{æ}$ the absorption coefficient, T_F the flame temperature and T_a the ambient temperature. For $d \gg 0.1$ m the heat conduction along the tank wall is no longer relevant. The convective term reaches its minimum at $d \approx 0.1$ m where is the minimum of $\overline{v}_a$. For $d > 1$ m the flame is optically thick, gray radiator. Here, is the radiative heat back to the liquid pool the dominant process.

The contributions from heat conduction, convection and radiation must be taken into account [32]. In the case of small pool diameters is the energy input on the pool edge or on the tank wall in comparison to the energy input on the pool surface crucial. This influence is inversely proportional to d, consequently, for the larger pools, the heat conduction is negligible. In the transition area is practically, only the convective transfer in this area, since the flame is optically still relatively thin [41]. With a growing of pool diameter (from d ≥ 1 m) the contribution of thermal radiation is dominant. The combustion process in turbulent diffusion flame is given on Fig. 2.5.

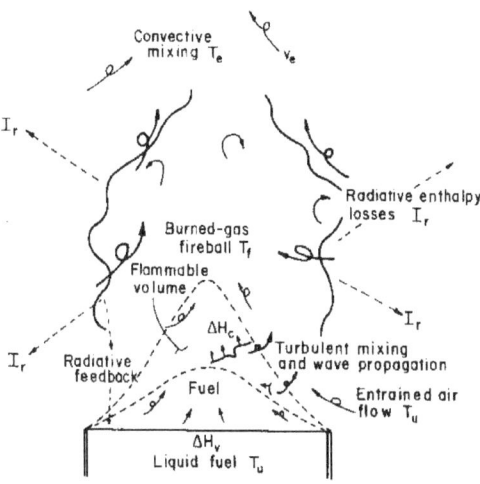

Fig. 2.5: Schematic illustration of combustion process in turbulent diffusion flame according to [34].

For d > 1 m, based on the above Eq. (2.1b) for burning of liquid fuels (e.g. methanol, butane, hexane and gasoline) Burgess [36] and Hetzberg [34] give a simplified relation:

$$\overline{v}_{a,max}(d) = 1.27 \cdot 10^{-6} \frac{\Delta h_c}{\Delta h_v} \tag{2.4}$$

where Δh_c and Δh_v are combustion and evaporation enthalpy.

For fuel mixtures, a uniform burning rate can not be set. At the beginning the lower boiling component i burns, while later when the mass fraction of the mixture increases the higher boiling components being heated to boiling point so than the burning rate of the flame can be determined. The maximal burning velocity $\overline{v}_{a,max}$ of liquid mixtures is given according to [37]:

$$\overline{v}_{a,max}(d) = 1.27 \cdot 10^{-6} \frac{\sum_i \tilde{y}_i(-\Delta h_{c,i})}{\sum_i \tilde{y}_i \Delta h_{v,i} + \sum_i \tilde{x}_i \int_{T_a}^{T_b} \overline{c}_p(T)dT} = \frac{\sum_i \tilde{y}_i(-\Delta h_{c,i})}{\Delta h_v^*} \quad . \quad (2.5a)$$

and for $(-\Delta h_{c,i}) = \Delta h_{v,i}$ and $\tilde{y}_i > \tilde{x}_i$ (e.g. for gasoline) [7]:

$$\overline{v}_{a,max} = \sum_i \tilde{y}_i \overline{v}_{a,i} \quad . \quad (2.5b)$$

Here $\tilde{x}_i$ and $\tilde{y}_i$ are the molar fractions of the liquid and gas phase which c_p is the specific heat capacity, T_a and T_b are atmospheric and boiling temperature. The denominator includes the dependence of combustion enthalpy on molar masses and on the temperature and in the above relation is used as Δh_v^*.

Quantitatively, mass burning rate and the combustion behaviour of combustible materials are described in the context of burning rate (or burning velocity v_a (m/s)) and mass burning rate in the time per unit mass of the fuel.

The time averaged mass burning rate $\overline{\dot{m}''_f}$ in kg/(m² s) can be calculated multiplying the time averaged burning velocity and a liquid density of fuel and gives:

$$\overline{\dot{m}''_f} = 10^{-3} \frac{\Delta h_c}{\Delta h_v^*} \quad (2.6)$$

which is valid for a wide range of gaseous and liquid fuels [7].

For the radiative optically thin and thick flame regimes Hottel [32] gives the following correlation:

$$\overline{\dot{m}''_f} = \frac{\sigma \overline{T}_{f,eff}^4 \left(1 - e^{-k\beta d}\right)}{\Delta h_v} \quad . \quad (2.7)$$

An empirical correlation for mass burning rate is given by Burgess [36]:

$$\overline{\dot{m}''_f} = \frac{10^{-3} \times (-\Delta h_c)}{c_p(T_b - T_a) + \Delta h_v}. \tag{2.8}$$

For the maximum mass burning rate the following equations are given [7,36]:

$$\overline{\dot{m}''_f}(d) = \overline{\dot{m}''}_{f,max}\left(1 - e^{-k\beta d}\right) = \overline{\dot{m}''}_{f,max}\,\varepsilon_F, \tag{2.9}$$

$$\overline{\dot{m}''}_{f,max} = \rho_f \overline{v}_{f,max} \approx 1.27 \times 10^{-6}(-\Delta h_c / \Delta h_v)\rho_f. \tag{2.10}$$

2.2 Flame geometry

The description of a flame depends mostly on the length of the flame diameter, flame length, the burning rate or mass burning rate the temperature and the flame radiative properties. These properties are usually taken as averaged in time. The measurements derived from different assessments for the influence factors and the geometry of large flames are shown below.

2.2.1 Flame length

Several important physical processes in tank and pool fires are shown in Fig. 2.1. The time averaged visible flame height $\overline{H}$ is defined as the height of the clear flame zone $\overline{H}_{cl}$ and the plume zone $\overline{H}_P$ [7] or the heights of the clear flame zone $\overline{H}_{cl}$, pulsation zone $\overline{H}_{pul}$ and the plume zone $\overline{H}_P$ (Fig. 2.6) [7], depending on the model. For the estimation of the thermal radiation of larger, sooty fires the length $\overline{H}_{cl}$ is of importance (Chapter 5.2).

The time averaged relative $\overline{H}/d$ and maximum relative $(\overline{H}/d)_{max}$ visible flame height may, dependent on a flame Froude number Fr_f and non dimensional wind velocity $\overline{u}_w^*$, which can be estimated with the following correlations:

$$\overline{H}/d = a\,Fr_f^b\,\overline{u}_w^{*c} \quad \text{and} \quad (H/d)_{max} = a\,Fr_f^b\,\overline{u}_w^{*c}. \tag{2.11a,b}$$

$$\overline{u}_w^* = \overline{u}_w / \overline{u}_c \text{ or } \overline{u}_w^*(10) \equiv \overline{u}_w(10\,\text{m}) / \overline{u}_c \tag{2.11c}$$

is a scaled wind velocity with $\overline{u}_c = (g\overline{\dot{m}''_f}d/\rho_v)^{1/3}$, (2.11d)

a,b and c are experimental parameters which detailed values can be found in [7].

12   2 Some characteristics and modeling of pool fires

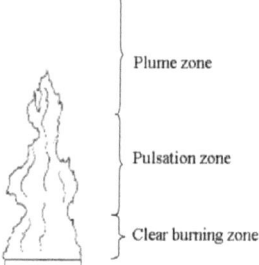

Fig 2.6: Three flame zone in a large pool fire.

There is a relatively large number of correlations (e.g. Thomas [42], Stewart [43], Moorhouse [44] and Heskestad [45]) which are often used, which have differing empirical parameters a, b, c, as given in Tab. 2.1.

Table 2.1: Parameters for determination of dimensionless visible flame lengths used in Eq. (2.11a,b).

Correlation	a	b	c	Comment
Thomas 1	42	0.61	0	Measured on wood fires without wind; $\bar{H}/d$; [42]
Thomas 2	55	0.67	−0.21	Measured on wood fires with wind; $(\bar{H}/d)_{max}$; [42]
Moorhouse	6.2	0.254	−0.044	Measured on large LNG pool fires; $(\bar{H}/d)_{max}$, $\bar{u}_w^* = \bar{u}_w^*(10)$ [44]
Muñoz 1	8.44	0.298	−0.126	Measured on gasoline and diesel pool fires; $(\bar{H}/d)_{max}$; [21]
Muñoz 2	7.74	0.375	−0.096	Measured on gasoline and diesel pool fires; $\bar{H}/d$; [21]
Muñoz 3	11.76	0.375	−0.096	Measured on gasoline and diesel pool fires; $(\bar{H}/d)_{max} = 1.52\,\bar{H}/d$; [21]

As shown in Fig. 2.7 the Thomas equation better match the experimental data. The visible flame length without wind influence according to Thomas [42] is predictable, with the parameters based on experiments with wood fires where in a case of calm conditions: a = 42, b = 0.61, $\overline{u}_w^{*c}$ = 1, and in a case of the wind influence: a = 55, b = 0.67, c = 0.21.

For an approximate assessment of the flame height of e.g. the Buncefield incident [1,2] the maximum, *visible, relative* flame height according to Eq. (2.11b) for gasoline fires is calculated, where c = 0 (no wind effect) is set:

$$(H/d)_{max} \approx a\ Fr_f^b = a \left(\frac{\overline{m}_f''}{\rho_a \sqrt{gd}} \right)^b . \qquad (2.12a)$$

In Fig. 2.7 is shown a relationship between the dimensionless burning rate and a relative flame length H/d [5].

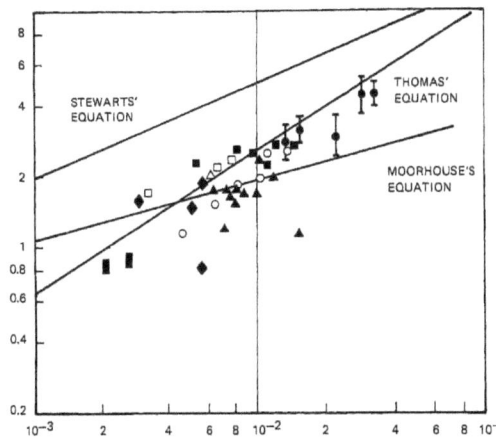

Fig. 2.7: Correlation between the relative length and the flames dimensionless rate of burning [5].

With $\overline{m}''_{f,max}$ (d ≥ 9 m) = 0.083 kg/(m²s) for a gasoline pool fire, ρ_a = 1.29 kg/m³ and the parameter a, b, from Tab. 2.1 an approximate calculation using Eq. (2.12a) gives:

$1.8 < (H/d)_{max,calc} < 1.9.$ (2.12b)

For the time averaged relative flame height $\overline{H}/d$ of a gasoline pool fire ($d \geq 9$ m) the calculation using Eq. (2.11b) and Tab. 2.1 approximates to:

$$(\overline{H}/d)_{calc} \approx a \; Fr_f^b = 7.74 \left(\frac{\overline{\dot{m}_f''}}{\rho_a \sqrt{gd}} \right)^{0.375} \approx 1.2.$$ (2.12c)

From measurements on relatively small gasoline pool fires ($1.5 \text{ m} \leq d \leq 6 \text{ m}$) [7] a value for the time averaged flame height $(\overline{H}/d)_{exp}$ was found:

$1 < (\overline{H}/d)_{exp} < 1.9.$ (2.13a)

For a relatively large gasoline tank flame (d = 23 m) the following value was measured [7]:

$(\overline{H}/d)_{exp} = 1.7,$ (2.13b)

where a time averaged flame height is assumed.

From the Eqs. (2.11b) and (2.12c) it follows, that the maximum relative flame height $(H/d)_{max,Bunc}$ in a case of a very large pool fire as occurred in the Buncefield incident is extraordinarily high in comparison to the calculated maximum relative flame heights $(H/d)_{max,calc}$ [1].

If the empiric relationship [1]:

$(H/d)_{max} = 1.52 \; \overline{H}/d$ (2.14a)

is also considered valid for the Buncefield tank fire, then it follows from Eq. (2.11b) an empiric relationship for the time averaged relative flame heights in the Buncefield incident:

$1.7 < (\overline{H}/d)_{exp,Bunc} < 4.3.$ (2.14b)

From the Eqs. (2.12b) and (2.14b) it therefore follows, that most probably, also the time averaged relative flame heights in the Buncefield incident are extraordinarily high, in comparison to the calculated and measured time averaged relative flame lengths $(\overline{H}/d)_{calc}$ and $(\overline{H}/d)_{exp}$ [1].

In the case of the influence of side winds for the calculation of flame length exists also the Stewart correlation [43] relates with the properties of liquid methane. As shown in Fig. 2.7, the flame lengths according to Stewart [43] are not in particularly good agreement with experiments. The Moorhouse relation [44] based on experiments with LNG pool fire shows particularly a good agreement for some fuels. The wind influence in Eq. (2.11a,b) is already taken into account. The calculation of flame length according to Heskestad [45] is based on experiments with hydrocarbon flames, without taking into account the wind effects.

The flame length is generally predicted as a maximum length or time averaged visible length [22]. The tractable by the human eye visible wavelength is around 380 nm $< \lambda <$ 750 nm. However, it is difficult to accurately determine visible flame lengths. Another definition is based on the contours of the stoichiometric composition in the flames [22].

The need for determination of a flame length in a safety context is correlated with the heat radiation. If in the determination of the relative flame length the cold soot particles are taken into account, the thermal radiation as integral of time and area averaged can reach only a low value. If only the visible part of the flame is used as the flame length value a correspondingly high heat radiation is adjacent. The flame length H is often not indicated as an absolute but relative value to the pool diameter d. As shown in [21,42-47] a derived flame length to diameter ratio (relative flame length) varies theoretically and experimentally determined from 0.2 to 4.5, depending on the pool diameter, wind influence and type of fuel. The flame length can be determined by means of the visible images obtained from the VHS video recordings.

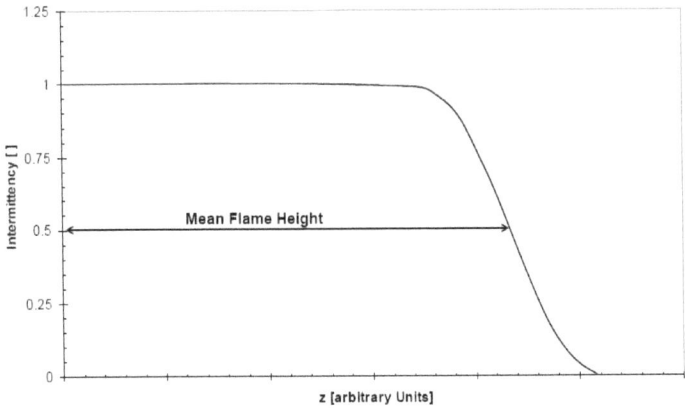

Fig. 2.8: Graph of intermittency vs. flame height [47].

The maximum height of the visible luminous flame can be selected from each frame in the sequence [21]. The continuous region should be comparable to the lowest flame height measured during experiments, and the intermittent region should be the area between this point and the maximum upper flame height measurement. The mean flame height has been defined by Zukoski and co-workers [46] in terms of the intermittency of the flame, I. The intermittency is defined as the fraction of time in which the flame reaches a certain height, z (m). The mean flame height is then the height at which the intermittency is 0.5 [21,46]. A typical graph of intermittency vs. flame height is given in Fig. 2.8 [47].

2.2.2 Flame tilt

Under the influence of the cross wind a flame tends to be tilted under a certain angle θ (Fig. 2.9).

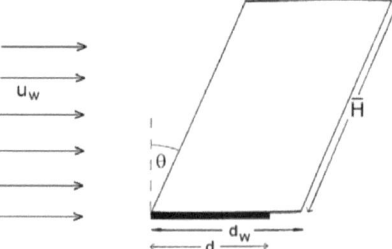

Fig. 2.9: Inclination angle θ of the flame under the wind influence.

Numerous laboratory studies have shown that the flames inclination can be calculated depending on the Froude and Reynolds number. In general, the inclination angle θ of flames can be calculated with equations of Welker and Sliepcevich [48], Thomas [42] and the American Gas Association (AGA) [49]. According to Welker and Sliepcevich [48] the inclination angle θ can be calculated as follows:

$$\frac{\tan\theta}{\cos\theta} = 3.3 \mathrm{Fr}^{0.07} \mathrm{Re}^{0.8} \left(\frac{\rho_v}{\rho_a}\right)^{-0.6}, \qquad (2.15a)$$

with $\mathrm{Re} = \dfrac{\overline{u}_w d}{u}$ and $\mathrm{Fr} = \dfrac{\overline{u}_w^2}{gd}$. \qquad (2.15b,c)

2.2 Flame geometry

The flames inclinations calculated based on the equations of Welker and Sliepcevich [48], however, show no good agreement with experimental data (LNG fire). The formula for calculating the inclination angle θ of flames according to Thomas [42] is based on experiments with wood fires:

$$\cos\theta = 0.7 \frac{\overline{u}_w}{(g\dot{m}_f'' d/\rho_a)^{1/3}}. \qquad (2.16a)$$

According to the AGA [49] is the tilt angle θ is calculated as follows:

$$\cos\theta = \begin{cases} 1 & \text{for } \overline{u}^* \leq 1 \\ 1/\sqrt{\overline{u}^*} & \text{for } \overline{u}^* \leq 1 \end{cases} \qquad (2.16b)$$

and the dimensionless wind velocity is

$$\overline{u}^* = \frac{\overline{u}_w}{(g\dot{m}_f'' d/\rho_a)^{1/3}}. \qquad (2.16c)$$

The wind measured at a height x = 1.6 m.

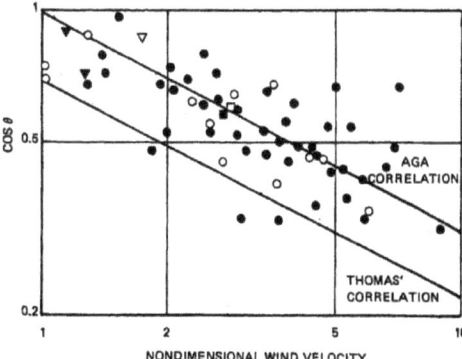

Fig. 2.10: Flames inclination θ of flammable liquids as a function of dimensionless wind velocity u_w.

Although the measured inclination flames are widely dispersed, the Fig 2.10 show that the flames inclination θ calculated by the AGA method [49] match better with experiments than those of Thomas [42].

2.2.3 Flame drag

Under a flame drag the extension of the flame base is assumed. As shown in Fig. 2.10 the flame drag d_w greatly depends on the wind velocity. Welker and Sliepcevich [48] give the dependence of flame drag on the Froude number for hydrocarbon fires:

$$\frac{d_w}{d} = 2.1 Fr^{0.21} \left(\frac{\rho_v}{\rho_a}\right)^{0.48}. \tag{2.17}$$

Moorhouse [44] gives the following dependence:

$$\frac{d_w}{d} = 1.5 Fr^{0.069}. \tag{2.18}$$

based on LNG pool fire experiments, with the wind speed measured the height x = 10 m. The flame drag according to Eq. (2.18) shows a good agreement with experimental data. For hydrocarbon pool fires, based on the Eq. (2.17) the following relationship can be used:

$$\frac{d_w}{d} = 1.25 Fr_{10}^{0.069} \left(\frac{\rho_v}{\rho_a}\right)^{0.48}. \tag{2.19}$$

Generally, at a rectangular pool the flame drag is clearly observed, the flame area and hence the heat radiation from the flame increases more than in a case of circular pool [19]. In the case of circular pool fires, under the wind influence the pool becomes more elliptical [19]. Consequently, the view factor of the flame on the receiver element surface changes due the flame drag.

2.3 Flow velocities

Often used techniques for determination of flow field, usually for pool flames with diameters d $\ll$ 1 m are Laser-Doppler Anemometry (LDA) [50] and Particle Image Velocimetry (PIV) [51-53]. In a case of both methods, the small particles are involved in flames and their speeds within the flame can be easily determined. It is assumed that the particle velocity is equal to the respective local flow velocity in the flame. In the LDA method the scattering of the moving particles changes their speed which causes frequency shifts in the received laser light (Doppler effect). In the PIV method, the particles are stimulated to illuminate with the energy of expanded laser beam, and their speed and direction can be determined by digital image analysis. Both methods are not practicable in large pool fires because of their dimensions and especially the usually very high density of soot particles which absorb a large part of the radiation. Through film recording of the VIS range of the flames and subsequent digital image analysis, the speed of coherent structures such as soot parcels (Section 2.5.3) on the flame surface can be determined [3,4,30,54-56]. The ascent speeds of these structures are usually not equating with the local reign the velocities, but qualitatively reflect only the velocities at the flames surfaces. In large pool flames, the flow velocities can be determined by measuring pressure difference. This method determines speeds only in a vertical direction. Velocity fields such can be determined by PIV, e.g. the flow in a horizontal direction can not identified. In general, in large pool flames with increasing pool diameter an increase in vertical velocities is recorded. Koseki [57] identified, for example, in n-heptane pool flames an increase of time averaged axial velocities at H/d = 1.5 from $\bar{u}$ = 3 m/s for d = 0.3 m to $\bar{u}$ = 17 m/s for d = 6 m (Fig. 2.11). The actual maximum value of $\bar{u}$ for d = 6 m was probably even higher, since in the experiments the entire amount of this flame could not be covered with probes and the actual maximum is outside the covered area. Koseki´s results show a dependence of the average vertical flow velocity $\bar{u}$ on the square root of the pool diameter d.

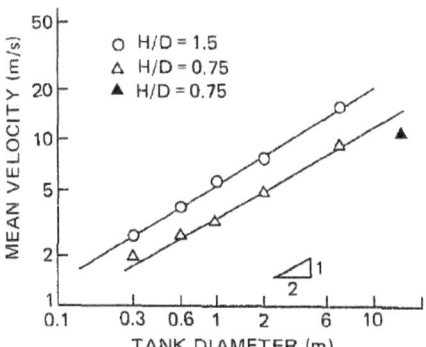

Fig. 2.11: Effect of tank diameter on mean velocity at H/d = 0.75 and 1.5 of n-heptane and JP-4 pool fire (solid triangle).

McCaffrey [47] compares both the temperatures and the axial flame velocity of pool fires with different heat release rates $\dot{Q}$ by showing the height as $\frac{x}{\dot{Q}^{2/5}}$ in dependence on normalized flow velocity as

$$\frac{u}{\dot{Q}^{1/5}} = k\left(\frac{x}{\dot{Q}^{2/5}}\right)^{\eta}. \tag{2.20}$$

2.4 Flame temperature

Flame temperature $\overline{T}$ is a function of pool diameter d, fuel f, area $\overline{a}_i$ of organized structures i (Chap.2.5) in the flame and effective absorption coefficient $\overline{\ae}_{eff,i}$: $\overline{T} = f(d, f, \overline{a}_i, \overline{\ae}_{eff,i})$ [1]. The calculation of the real flame temperature $\overline{T} < \overline{T}_{adiabat}$ is limited by large uncertainties [1].

According to [1,3,4,7] the flame temperature can be presented as time averaged temperatures of organized structures in a flame e.g. for large, sooty, hydrocarbon pool fires as JP-4 pool fire these temperatures are: $\overline{T}_{re} \approx 1413$ K, $\overline{T}_{hs} \approx 1329$ K, $\overline{T}_{sp} \approx 623$ K.

From thermographic measurements [3,4] logarithmic-normal distribution of the flame temperature log-normal pdf $g_T(\overline{T})$ as f(d,f) can be determined (Chap. 2.6.2).

The measurement of temperatures inside of pool flames can be directly done with thermocouples or indirectly through the radiation measurements such as IR-thermographic system (Chap. 3.3) or with radiometers [3,4]. The different methods offer different advantages and disadvantages. The measurement with thermal elements can in principle be done on a variety of locations within the flame offering the temperature profiles in the horizontal (radial) and vertical (axial) direction. By building a large number of probes can not carry the flame undisturbed any more. For pool flames with very large diameters and correspondingly large flame lengths is the realization of such a measurement setup over the entire flames very expensive. In those flames thermostats are therefore usually used only to measure the temperatures in the lower and middle areas of the flame. When using thermocouples is not always ensured that the temperature of the probes is equal to the surrounding gas. Temperature differences may be due to cool the probes, radiation, heat conduction and heating catalytic reactions related to the probe surface. At very high velocities such as in jet flames can be an additional aerodynamic heating presented. Too slow thermocouples can also not record fast temperature changes. Planas-Cuchi and Casal [58] determine, for example, the maximum flame temperature of a hexane pool flame at a pool surface $A_P = 4$ m^2 with thermocouples to be $T_{max} = 957$ K while Bainbridge [59] indicates $T_{max} = 1150$ K. Planas-Cuchi and Casal explain the difference by the slowdown in the probe because of the radiated energy and they came in line with Gregory [60] to the conclusion that thermocouples due to this effect provide readings in this generally lower flame temperature. Temperature measurements in flame generally do not affect radiation measurements. A disadvantage is that, especially for large pool flames, which produce a lot of smoke and thus are optically thick, only the radiation of the flame surface can be registered. The radiation from inside the flames is blocked by absorption of a dense soot parcels. In contrast to the specific measurements with thermocouples with the same IR thermographic system, the spatial resolution is device specific. The temperature determination by radiometer measurements can vary depending on the covered section of the flame and for only mean relatively large parts of the flame surface or even just for the whole flame may be indicated. To the temperature from radiation measurements various sizes must be known. For the radiation received at the receiver it is applied:

$$I_F = \varepsilon_F \tau_a \sigma \left(T_F^4 - T_a^4 \right). \tag{2.21}$$

The transmittance τ_a in the air depends on the humidity and other gas components such as CO_2 [59]. The emission of large flames degrees (d > 1 m), is in most cases to $\varepsilon_F = 1$ adopted. ε_F is basically dependent on the pool diameter and the type of fuel. Planas-Cuchi et al. [58] based experiments on gasoline and diesel flames with a pool diameter 0.13 m ≤ d ≤ 0.5 m, show that in this interval with decreasing pool diameter emission is being significantly reduced. The emissivity seems to increases by increasing d to approach approximated $\varepsilon_F = 1$. The influence of the fuel is in their experiments so low that can be neglected. The temperatures within a flame generally depend on many factors, such as a fuel, a pool diameter and also wind conditions. The maximum amount of heat released from a flame is given by

$$\dot{Q} = \dot{m}_f'' \Delta h_c A_P. \tag{2.22}$$

2 Some characteristics and modeling of pool fires

According to McCaffrey [47], the temperature and the velocity profiles (Chap. 2.3) along the axis of flames with different values for $\dot{Q}$ are scaled and the values for heights $\dfrac{x}{\dot{Q}^{2/5}}$ and velocity $\dfrac{u}{\dot{Q}^{1/5}}$ are normalized. The factors $\dot{Q}^{2/5}$ and $\dot{Q}^{1/5}$ are purely empirical values derived from his experiments with a gas burner with a square pool of 0.3 m. The heat release rates $\dot{Q}$ depend on the variation of the fuel flow. McCaffrey divides the flame into regions in and over the flame, to lower, clear flame zone, the transition zone and the flame plume (Fig. 2.3).

Fig 2.12a,b shows the Koseki´s [57] by thermocouples measured axial and radial temperature distribution within a heptane pool fire with d = 6 m as an example which may be applied to other heptane pool flames with diameters 0.3 m ≤ d ≤ 6 m. Since the bottom of unburned and relatively cold fuel vapors rising from the pool, has a significantly lower temperature than in a larger dimensionless heights about $0.6 \leq x/r \leq 1.7$. With increasing radial distance the average temperatures decrease. The dependence of the time averaged axial temperatures of the flames on the pool diameter and the fame height is shown on the Fig. 2.12b.

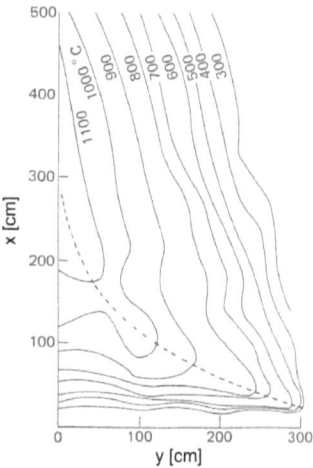

Fig. 2.12a: Isotherms of n-heptane pool fire (d = 6 m) according to [57].

The flame temperatures are uncorrected, means that the heat loss in thermocouples by radiation was not considered [57]. The mean temperature increases with increasing pool diameter up to T_{max} = 1473 K for d = 6 m. An exception is only the smallest flame with d = 0.3 m, with slightly higher temperatures than the flame with d = 0.6 m. The dependence of maximum average temperature of all flames except for the little ones, on dimensionless height is $1.3 \leq x/r \leq 1.6$. Similar temperature profiles flames in the axial and a radial direction are also measured in [58] for gasoline and diesel pool flames with $1.5 \text{ m} \leq d \leq 4 \text{ m}$. The mean maximum temperature of T_{max} = 1223 K for $x/r \approx 0.25$, and is significantly lower than in Koseki's experiments.

While Koseki registered a temperature near maximum at a relative height of $x/r = 2$, in [58] at the same relative height is shown the significant decrease to $T \approx 573$ K. The difference may be in the physical properties of the various fuels used. While gasoline and heptane have $\Delta h_{c,gasoline}$ = 43 MJ/kg and $\Delta h_{c,heptane}$ = 44.7 MJ/kg, the almost identical enthalpy of combustion they differ in their mass burning rates $\overline{\dot{m}_f''}_{gasoline}$ = 0.05 kg/(m² s) and $\overline{\dot{m}_f''}_{heptane}$ = 0,101 kg/(m² s) for almost a factor 2.

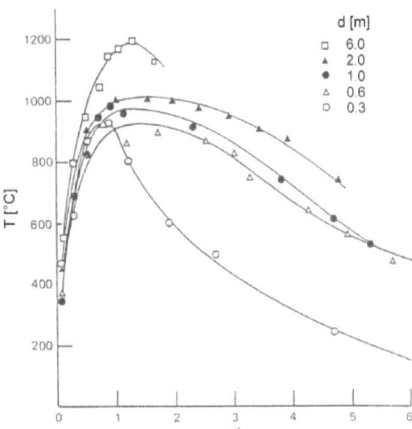

Fig. 2.12b: Time averaged axial temperature profiles of n-heptane pool fires depending on d [57].

2.5 Organized structures

With pool diameters d > 1 m different types of structures in a fire can be classified (Fig. 2.13), differing mainly in their temperature and thermal radiation. Some models to calculate the thermal radiation take into account the radiative properties of different structures (Chapter 2.6.2), so these differences in the following sections are highlighted.

Fig. 2.13: Organized structures in a large, sooty, hydrocarbon pool fire [61].

2.5.1 Reactive flame zone

An effective reaction zone (re) is defined as a very hot (T_{re}) emitting and self-absorbing homogeneous volume of flame gases and soot particles. It is assumed that the thermal radiation of the pool fire originates from these reaction zones which have an time averaged effective length-scale $\bar{l}_{re}(d)$ and the temperature $\bar{T}_{re}$. By using a simplified radiation transport equation for an absorbing and emitting soot particle/flame gas mixture it is shown that, in that case the band lines of the gases are negligible, the time averaged radiant emittance $\overline{M}_{re}(d)$ of an approximately grey emitting effective reaction zone $i = re$ can be calculated from the equations [3,7]:

$$\overline{M}_{re}(d) = \left(1 - \hat{\tau}_{re}(d)\right)\sigma\left(\overline{T}_{re}^{4} - T_{a}^{4}\right), \tag{2.23a}$$

$$1 - \hat{\tau}_i(d) = \hat{\varepsilon}_{eff,i}(d) = 1 - e^{-k_{eff,i}(d)d}, \tag{2.23b}$$

$$\bar{k}_{eff,i}(T_i) = m_i \bar{\mathfrak{a}}_{eff,i}(T_i) = m_i B_i c_{s,i} \bar{T}_i, \tag{2.23c}$$

$$\bar{æ}_{\text{eff,i}}(T_i) = a_1 \bar{æ}_{\text{m,i}}(T_i),\tag{2.23d}$$

$$\bar{l}_i(d) = m_i d,\tag{2.23e}$$

where $\hat{\tau}_i$, $\hat{\varepsilon}_{\text{eff,i}}$, $k_{\text{eff,i}}$, $æ_{\text{eff,i}}$ are the modified, effective: transmittance, emissivity, absorption coefficient (m^{-1}), total grey absorption coefficient (m^{-1}); B_i, $c_{s,i}$, a_1, m_i are a factor (m^{2}kg^{-1}K^{-1}) [3,7], soot mass concentration (kg m^{-3}), a constant parameter [3,7], a parameter; σ, T_a are the Stefan-Boltzmann constant (Wm^{-2}K^{-4}), ambient temperature (K). In Eqs. (2.23b-e) the subscript i refers to the organized structures i = re, hs, sp. With respect to Eq. (2.23a), i = re.

Combustion zone (clear zone, luminous zone) is also sometimes known as a flame base and it is localized directly above the fuel surface (Fig. 2.13). There exists a good mixture of fuel vapor and atmospheric oxygen. The favorable ratio of oxygen to fuel results in high response rates, so that a large part of the combustion enthalpy is here released. Therefore, in the clear burning zone exist permanently high temperatures in the range of 873 K $\leq \bar{T}_{cl} \leq$ 1413 K and there is a high surface emissive power 33 kW/m$^2 \leq \overline{SEP}_{cl} \leq$ 430 kW/m^2 according to [3,4,7]. The clear burning zone seems to be viewed as a luminous ring around the base of the flame, its height $\bar{h}_{cl}$ varies in sooty flames depending on the fuel between 0.1d and 0.3d [7].

2.5.2 Hot spots

A hot spot (hs) is defined as an intensively emitting, absorbing and transmitting hot (T$_{hs}$) homogeneous volume of flame gases and soot particles which moves radial to the flame surface and surrounds the reaction zone. Hot spot is a structure with very high surface emissive power. Hot spots occur outside a visible area of high temperature and radiation intensity just above the combustion zone. It arises, for example, when hot combustion gases from the flames inner, because of their high speed reach through the sooty stains or postponement of the soot parcels, the outer flame area. Hot spots are like the soot parcels, a mixture of gases and particular emissions of flame, but their temperatures are much higher 873 K $\leq \bar{T}_{hs} \leq$ 1413 K [3,4,7] equal to those of the clear combustion zone. Due to the high temperatures these structures develop high buoyancy forces with speeds of $\bar{u}_{hs,max}$ = 20 m/s. It is assumed that each hot spot with an effective length scale $\bar{l}_{hs}$ (d) and a temperature $\bar{T}_{hs}$ partly transmits, absorbs and emits a large amount the incident radiant emittance $\overline{M}_{re}$ (d). The approximately grey radiation of the hot spot leads to a surface emissive power $\overline{SEP}_{hs}$ (d) which can be calculated from the equation [3,7]:

$$\overline{SEP}_{hs}(d) = \left(1 - \hat{\tau}_{hs}(d)\right)\sigma\left(\overline{T}_{hs}^4 - T_u^4\right) + \hat{\tau}_{hs}(d)\left(1 - \hat{\tau}_{re}(d)\right)\sigma\left(\overline{T}_{re}^4 - T_u^4\right). \qquad (2.24)$$

With respect to Eq. (2.24) the relationships in Eqs. (2.23b-e) are valid for $i = hs$. Due to their high surface emissive power 33 kW/m² $\leq \overline{SEP}_{hs} \leq$ 430 kW/m² can, however, by hot spots, short maxima radiation be harmful to humans. The $\overline{SEP}_{hs}$ presents the main part of the total surface emissive power $\overline{SEP}_{tot}$ (d) of pool fire [7].

2.5.3 Soot parcels

Soot parcels appear as gray to black vortex structures. The proportion of soot parcels in flames at the surface depends greatly on fuel f and the pool diameter d. There are flames in which, as in the case of LNG or peroxide pool flame [31] a very low amount of soot parcels occur. In large pool flames typical fuels such as gasoline, kerosene or diesel the most of the flame area A_F consists of soot parcels (Fig. 2.13).

In such heavy sooty flame begins its origin directly above the clear burning zone. Due to the cooling of the hot soot particles by the surrounding air, at the flame surface a layer of relatively cold, non-luminous soot particles exists. The temperature of soot parcels was investigated by Göck [3,4] with the help of an infrared thermography system, and determined as 523 K $\leq \bar{T}_{sp} \leq$ 873 K. Soot parcels are highly absorbent structures. As a solid state, absorb and emit the soot particles on the continuous spectrum. A part of the absorbed radiant energy is converted into heat energy and leads to a small increase of the soot temperature. The remainder of the previously absorbed radiation energy is re-emitted. A portion of the emitted radiation from flame inner is blocked by the soot parcels. In comparison to the clear burning zone (Section 2.5.1) and the hot spots (Section 2.5.2) have relatively weak soot parcels surface emissive power in the range of 33 kW/m² $\leq \overline{SEP}_{sp} \leq$ 50 kW/m² [3,4,7]. In the case of the most hydrocarbon pool fires with increasing pool diameter d soot fraction increases and the effect of smoke blockage effect of radiation increases, which results in decrease of an area-related emission of the flame.

A soot parcels (sp) with an effective length-scale $\bar{l}_{sp}(d)$ are defined as a strongly absorbing, relatively weakly emitting and transmitting, less hot ($\bar{T}_{sp}$) homogeneous volume of flame gases. A large amount of non-luminous soot particles are formed at the flame surface and surround the reaction zone. A large fraction of the absorbed exitance $\overline{M}_{re}$ by the soot particles will be transformed to non-radiant energy. Due to this smoke blockage effect the temperature of the large number of relatively cold soot particles will increase by a few degrees Celsius. The approximately grey radiation of the soot parcel leads to a surface emissive power $\overline{SEP}_{sp}(d)$ which can be calculated from the equation [3,4,7]

$$\overline{SEP}_{sp}(d) = \hat{\tau}_{sp}(d)\left(1-\hat{\tau}_{re}(d)\right)\sigma\left(\overline{T}_{re}^4 - T_a^4\right) + \left(1-\hat{\tau}_{sp}(d)\right)\sigma\left(\overline{T}_{sp}^4 - T_a^4\right). \quad (2.25)$$

With respect to Eq. (2.25) the relationships in Eqs. (2.23b-d) are valid for i = sp. The calculation of $\overline{SEP}_{sp}(d)$ shows that $\overline{SEP}_{sp}(d)$ is very low due to the smoke blockage effect of the fire.

To calculate the burning rates $\bar{v}_a(d)$ with an equation given in [3,7] the existence of fuel parcels (unburned fuel vapor) above the liquid fuel surface is additionally assumed.

In areas where exists an inadequate mixing of fuel and oxidizers, next to gaseous species soot is formed as a combustion product. The soot particles come as the spotlight especially in view of the radiant heat of the flame of particular importance. The term soot, however, includes a wide range of particles that are not identical chemical structure and have therefore some distinct characteristics. These differences are due to the different processes during the soot formation. Until today there is no complete understanding of soot formation all the underlying processes or reactions. Unity reigns but the general conduct of soot formation, it will appear the following processes:

- nucleation emergence of primary particles surface
- oxidation
- coagulation growth in the gas phase will begin to split the fuel into smaller molecules or radicals. The various intermediate products may react with each other

and form polycyclic aromatic hydrocarbons (PAH). The PAH are considered as precursors. The formation mechanism of PAH may vary depending on the fuel, different reactions [62] take place as (a) HACA (hydrogen abstraction, carbon addition) mechanism in the planar PAH growth, and (b) enlargement on the surface – growth of soot particles of liquid fuels, however, always consist of the acetylene. The ring structure and the subsequent growth ring can be regarded as a repeated secession by adding hydrogen and acetylene and will therefore come to a HACA mechanism [63]. With continuing the process in this way precursors composed of larger two-ring systems are formed [64]. Besides the addition of acetylene, also adding another already formed ring systems is happening. This coagulation is considered as a crucial step in the transition from the primary precursor particles [65] which continue with the aging and the surface growth of the particles, soot oxidation and continue to the coagulation. The surface growth can happen as the addition of acetylene [63]. This continually processes lead to a significantly decreasing surface activity of soot particles [66]. The soot oxidation can always run parallel to soot formation. As a key soot oxidation the process takes place at high temperatures and high oxygen partial pressures [67] reach its maximum in oxidation rate in ambient air at about 2000 K [68]. In addition to the O_2 molecule can also O radicals and especially OH radicals contribute to oxidation [68-70]. By oxidation, the resulting soot in appropriate conditions almost completely back into carbon dioxide and water. The properties of old soot particles from younger differ. To have precursors and primer particles have a relatively high proportion of hydrogen, which decreases with increasing aging.

2.6 Thermal radiation models

To describe the heat radiation of large pool flames different stationary, semi-empirical models have been developed, divided in two categories. The majority of the models describe the surface of the flame as an average specific area $\overline{A}_F$. The different calculation methods are described below.

2.6.1 Semi-empirical radiation models

The semi-empirical models are widespread used. There are relatively simple, often stationary models, which essentially refer to the flame geometry and predict the thermal radiation from the pool/tank, spill fires, flares, jet flames, fire balls and fire clouds, generally with more empirical parameters [7].

- The zone models are based on the differential equations for the conservation of mass and energy. The flame is divided into 2 to 20 certain zones. The computing times are usually short [7].

- The field models are generally stationary and based on a solving of time-averaged Navier-Stokes differential equations (partial differential equations) with often empirical sub models. These models (so-called cold models) are used mostly for predicting not reacting flows. Because of their mathematical complexity, these models, however, require large computing times [7].

- The integral models represent a compromise between the semi-empirical models and field models. These models are based initially on the same differential equations as the field models, however, include sub-models for turbulence, combustion reactions and heat transfer processes. Following simplistic assumptions in reducing the partial differential equations to ordinary differential equations the computing times significantly becomes smaller than in the case of the field models. Still, there is no integral model able to give an adequate prediction of consequences from accidental fire [7].

2.6.1.1 Point source radiation model PSM (Point Source Model)

The point sources radiation model (PSM) describes the thermal radiation received by an object under the assumption that the flame can be viewed as a point heat source (Fig. 2.14) [25]. The point source radiation model calculates the mean irradiance (thermal radiation flux) from the following relationships [7,25]:

$$\overline{E}_{PSM} = \frac{\overline{f}_{rad}\overline{Q}_{rad}}{4a\pi\Delta y^2} = \frac{\overline{f}_{rad}\overline{m}''_f A_P(-\Delta h_c)}{4a\pi\Delta y^2} \qquad (2.26)$$

with $A_P = r^2\pi$.

The total energy released $\overline{Q}_{rad}$ known and the radiation intensity is inversely proportional to the square Δy^2 the distance between the flame and the irradiated area element A_E. The mean irradiance on concentric circles with the radius Δy is then [7]:

$$\overline{Q}_{rad} = \overline{f}_{rad}\overline{Q}_c = \overline{f}_{rad}A_P(-\Delta h_c)\overline{m}''_f = \overline{f}_{rad}A_P(-\Delta h_c)\overline{m}''_{f,max}\left(1-e^{-k\beta d}\right). \qquad (2.27)$$

With the PSM the mean irradiance $\overline{E}_{PSM}$ may be calculated according to [7,25]:

30 2 Some characteristics and modeling of pool fires

$$\bar{E}_{PSM}(\Delta y / d) = \frac{\bar{f}_{rad}(-\Delta h_c) \bar{m}''_{f,max}(1-e^{-k\beta d})}{16(\Delta y / d)^2}, \quad \text{for } \Delta y/d > 4. \quad (2.28a)$$

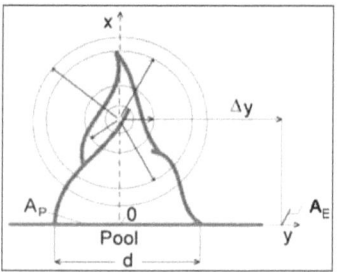

Fig. 2.14: Point source radiation model [7]

For the near field $\bar{E}_{PSM}$ is calculated according to:

for $0.5 < \Delta y/d < 4$

$$\bar{E}_{PSM}(\Delta y/d) = j_N \cdot 0.131 \, \bar{f}_{rad}(-\Delta h_c) \bar{m}''_{f,max}(1-e^{-k\beta d}), \quad (2.28b)$$

with $\varphi_N = \dfrac{\bar{H}/d}{\pi(\Delta y/d)^2}$. $\quad (2.28c)$

The PSM however has only a very limited range of validity and in particular in the near field great uncertainties exist.

The pool surface is $A_P = \pi r^2$. The empirical factor is usually assumed as 1.

A view factor $\bar{f}_{rad}$ dependent on fuel and the pool diameter according to Moorhouse and Pritchard [44] is calculated as:

$$\bar{f}_{rad} = \frac{\overline{SEP}}{\Delta h_c \bar{m}''_f}\left(1 + 4\frac{\bar{H}}{d}\right). \quad (2.29)$$

The point source model can predict radiation in larger distances from the flame with satisfactory results for the irradiance, and in closer distance is useless because it underestimated the thermal radiation. The reason lies in the adoption of a point source of radiation, because in closer distance from the flame the irradiance depends strongly on the length, shape of the flame and its orientation to the radiation.

2.6.1.2 Solid flame radiation models (SFM, MSFM)

Conventional cylindrical flame radiation model SFM (Solid Flame Model)

According to SFM (a single zone-radiation model without a black soot zone) the time averaged maximum (index ma) surface emissive power $\overline{SEP}_{SFM}^{ma}$ is calculated according to [7,25]:

$$\overline{SEP}_{SFM}^{ma} = \varepsilon_F \sigma (\overline{T}^4 - \overline{T}_a^4). \tag{2.30a}$$

With $\varepsilon_F = 0.95$ (grey flame), $\overline{T} = 1173$ K (900°C) follows from Eq. (2.30a) a constant, surface emissive power:

$$\overline{SEP}_{SFM}^{ma} = 100 \text{ kW/m}^2 \neq f(d,f). \tag{2.30b}$$

The SFM was often used until now for (apparently conservative) predictions, although the Eqs (2.30a,b) does not agree with the newer measurements [3,4] of $\overline{SEP}_{act}(d,f)$. Conventional cylinder flame radiation model (SFM) [7,14,25] is a kind of a radiation zone model (without soot zone) applied for the maximum surface emissive power of a specific pool or tank fire.

With the assumptions $\varepsilon_F = 0.95$ (i.e. gray flame) and (900° C) follows from Eq. (2.30a) the constant average $\overline{SEP}$. The flame emission ε_F size is very difficult to estimate, since it consists of emissivities of the products of combustion, soot, water vapor, CO_2 and it depends on the path length l of the fire and the wavelength λ. Only for larger pool fire, in the case of most hydrocarbons optically thick fires (d ≈ 3 m) a good approximation is $\varepsilon_F = 1$, although such a pool fire in principle is not a black radiator.

The flame temperature is both experimentally as well as theoretically difficult to determine, especially because the flame temperature of the flame surface is not homogeneous and the flame generally is not a black radiator. The average surface emissive power $\overline{SEP}$ is a typical "derived" size, experimentally (Eq. (2.30a)) is only fairly difficult to determine, especially due to the dependence on the view factor φ acc. (2.28b) and consequently dependent on the flame surface A_F or flame length H which are difficult to measure. This means that the numerical value of the critical area used by the A_F or the length H is demanding in determination of $\overline{SEP}$. From this comments it follows that Eq. (2.30a), based on the Stefan-Boltzmann law by determination of path of the radiation exchange between a flame (T) and environment

(T_a) is used for determination of $\overline{SEP}_{SFM}^{ma}(d,f)$ which is in praxis used only for limited purposes. The SFM is often used so far for conservative predictions, although the Eq. (2.30a,b) may not agree with recent measurements of $\overline{SEP}$ [3,4,7].

Modified cylindrical flame radiation model MSFM (Modified Solid Flame Models)
For the MSFM (in principal a single zone radiation model) the time averaged, maximum surface emissive power $\overline{SEP}_{SFM}^{ma}(d,f)$ is generally calculated according to [25]:

$$\overline{SEP}_{SFM}^{ma}(d,f) = \frac{\overline{f}_{rad}(d,f)\,\overline{m}''_f\,(-\Delta h_c)}{4\,\overline{H}(d)/d}. \tag{2.31}$$

A variation of the MSFM is a *two* zone radiation model with *a lower* clear burning zone (LZ) with $\overline{SEP}_{cl}^{ma}$ and an *upper* black soot zone (SZ) with $\overline{SEP}_u$. The two zones may be calculated with, for example in [72]:

$$\overline{SEP}_{cl}^{ma} = \overline{SEP}_{max}(1-e^{-kd}), \quad \text{and} \tag{2.32a}$$

$$\overline{SEP}_u = (1-\overline{a}_{SZ})\overline{SEP}_{cl}^{ma} + \overline{a}_{SZ}\overline{SEP}_{SZ}. \tag{2.32b}$$

For e.g. *gasoline*-pool fires with $\overline{a}_{SZ} = 0.98$, $\overline{SEP}_{max} = 130$ kW/m^2 and $\overline{SEP}_{SZ} = 20$ kW/m^2 it follows approximately from Eqs. (2.32a,b) with $k \approx 2.0$:

$$\overline{SEP}_{cl}^{ma} \approx 130\,(1-e^{-2d}) \approx 130 \text{ kW/m}^2, \quad \text{and} \tag{2.33a}$$

$$\overline{SEP}_u \approx 0.02 \cdot 130 \text{ kW/m}^2 + 0.98 \cdot 20 \text{ kW/m}^2 \approx 22.2 \text{ kW/m}^2. \tag{2.33b}$$

Conventional and modified cylinder flame radiation model (SFM, MSFM) are very widespread models for calculating the average emissive power [14,24,72]. In these models the flame is assumed to have a radiative cylinder area (Fig. 2.15). The cylinder diameter is equal to the pool diameter d, the cylinder is equal to the amount of medium flame length H.

The model shows the experimentally found dependences on fuel and on the pool diameter, but provides a rough estimation of the maximum SEP. It is currently used for conservative forecasts.

2.6 Thermal radiation models

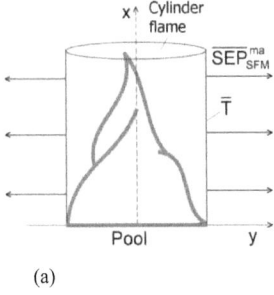
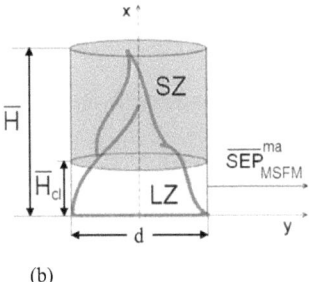

(a)   (b)

Fig. 2.15: Cylindrical flame model: (a) SFM: the flame is equally radiant cylinder, (b) MSFM: the flame is dividend into a clear luminous zone with a high radiation (LZ) and a non-radiating soot zone (SZ).

Generally, the average $\overline{SEP}$ of a pool fire depending on pool diameter is calculated as follows:

$$\overline{SEP}(d, f) = \overline{f}_{rad}(d)\, \overline{SEP}_{theor}, \text{ with} \tag{2.34a}$$

$$\overline{SEP}_{theor} = \overline{\dot{m}}_f''(\Delta h_c)\frac{A_P}{A_F} \equiv \overline{\dot{q}}_f'' \frac{A_P}{A_F}. \tag{2.34b}$$

For the cylinder flame area is given by:

$$\overline{A}_F = \pi d \overline{H}(d) + \frac{\pi d^2}{4}. \tag{2.34c}$$

Modified cylinder flame - radiation model (MSFM) generally calculates the maximum time averaged $\overline{SEP}$ of a pool or tank fire, depending on the diameter d and the type of fuel f according to [25] and from the Eq. (2.34b) follows that $\overline{SEP}$ is difficult to measure, especially due to dependence on the flame surface A_F or dimensionless flame length $\overline{H}/d$ which measurements have very unsafe levels.

The larger pool fire is not really the theoretical radiator alone emitting SEP from its surface. Rather, the pool fire can be seen as a volume emitter [73], i.e. the thermal radiation emitted varies with the path length on the issue. Therefore, the use of SEP two-dimensional or one-dimensional simplification or approximation of a very complex three-dimensional radiant heat phenomenon is to rough approximation [74]. Actually originating from the outer surface flames emit thermal radiation from gaseous combustion products (especially H_2O and CO_2 vapor), hot fuel vapor and particles, especially glowing soot particles (at sooty, larger fires) that exist in a certain depth in the interior of the flame.

2.6.1.3 Two zone radiation models (TZM)

Radiation model according to Mudan

Corresponding to the empirical radiation model according to Mudan [5] for sooty pool fires and the time averaged surface emissive power the following is valid (Fig. 2.16):

$$\overline{SEP}_{act}(d) = \overline{SEP}_{LS}^{ma}\, \overline{a}_{LS}(d) + \overline{SEP}_{SA}\,(1-\overline{a}_{LS}(d)) \tag{2.35a}$$

or with the area fractions

$$\overline{a}_{LS}(d) = \overline{A}_{LS}/\overline{A}_F = 1 - \overline{a}_{SA} = e^{-sd} = e^{-0.12d}, \tag{2.35b}$$

$$\overline{SEP}_{act}(d) = 140\, e^{-0.12d} + 20\,(1-e^{-0.12d}). \tag{2.35c}$$

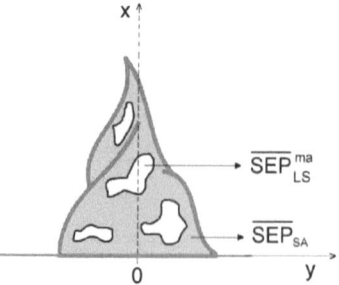

Fig. 2.16: Two zone radiation model [7]

This means $\overline{SEP}_{LS}^{ma} = 140$ kW/m² ≠ f(d,f) as well as $\overline{SEP}_{SA} = 20$ kW/m² ≠ f(d,f). Also $\overline{SEP}_{act}(d) \approx 20\,(1-e^{-0.12d})$ is valid for d ≥ 20 m, so *that for larger* pool fires the hot, luminous spots (Term 1 in Eqs (2.35a,c)) are eliminated.

In [21] instead of the Eqs (2.35b,c) there exists other relations for example $\overline{a}_{LS}(d) = 1.80\,d^{-0.377} - 0.533$ (for gasoline fires, d ≥ 5 m), $\overline{SEP}_{LS}^{ma} = 115$ kW/m² (for gasoline and diesel fires, d ≥ 5 m) as well as $\overline{SEP}_{SA} = 40$ kW/m² ≠ f(d,f).

2.6.2 Organized structures radiation models (OSRAMO II, OSRAMO III)

OSRAMO II

Organized structures radiation models OSRAMO II and stochastic OSRAMO III in Schoenbucher et al. [1,3,4,7,26] are based on also at the first time experimentally found coherent structures in pool flames (Chapter 2.5).

The Organized Structure Radiation Models (OSRAMO) taken into account the specific $\overline{SEP}$ of hot spots (hs) and soot parcels (sp). It is assumed that the hot spots, soot parcels, effective reaction zone (re) and the fuel bales (fp) have homogeneous entity characterized by the lengths l_i (i = hs, sp, re, fp). These organized (dissipative) structures i can partially emit, absorb and transmit a thermal radiation. It is further assumed that these structures i have different, but constant medium-modified temperatures and effective absorption coefficient. It is also assumed that the hot spots and soot parcels with diameters dependent area share occurrence on the flames surface. In the models OSRAMO II, III will be the first time the highly complex three-dimensional thermal radiation phenomenon adequately taken into account.

The thermal radiation is by the hot spots and soot parcels absorbed and then partially re-emitted. The average specific $\overline{SEP}$ of the entire surface according to OSRAMO II consists of the $\overline{SEP}$ of the structural elements: soot parcels and hot spots (Fig. 2.17):

$$\overline{SEP}_{OS}(d) = \overline{SEP}_{hs}\,\overline{a}_{hs}(d) + \overline{SEP}_{sp}\,\overline{a}_{sp}(d). \qquad (2.36)$$

$\overline{SEP}_{hs}^{ma}(d,f)$

$\overline{SEP}_{sp}(d,f)$

$\overline{H}_{cl}$

d = 25 m

Fig. 2.17: Organized structures in large JP-4 pool fire [1,7]

2 Some characteristics and modeling of pool fires

With OSRAMO II the time averaged surface emissive power $\overline{SEP}^{\parallel}_{act}(d)$ for larger, sooty pool fires is calculated based on experimental data of large JP-4 pool fires [1,3,4,7]:

$$\overline{SEP}^{\parallel}_{act}(d) = \overline{a}_{hs}(d)\, \overline{SEP}^{ma}_{hs}(d) + \overline{a}_{sp}(d)\, \overline{SEP}_{sp}(d), \tag{2.37a}$$

with $\overline{SEP}_i(d)$ for the dissipative structures i = hs, sp:

$$\overline{SEP}_i(d) = (1 - \overline{\tau}_i(d))\sigma(\overline{T}_i^4 - \overline{T}_a^4) + \overline{\tau}_i(d)(1 - \overline{\tau}_{re}(d))\sigma(\overline{T}_{re}^4 - \overline{T}_a^4), \tag{2.37b}$$

the *modified transmissivities* of the dissipative structures i = re, hs, sp:

$$1 - \overline{\tau}_i(d) = \overline{\tilde{\varepsilon}}_{eff,i} = 1 - \exp(-\overline{\tilde{\mathit{æ}}}_{eff,i} d), \tag{2.37c}$$

the *modified absorption coefficients* of the dissipative structures i = re, hs, sp:

$$\overline{\tilde{\mathit{æ}}}_{eff,i}(T) = \overline{\mathit{æ}}_{eff,i}\, b_i = 1.81 \cdot 10^3\, \overline{f}_v b_i \overline{T}_i \approx 1.12 \times 10^{-3} b_i \overline{T}_i, \tag{2.37d}$$

the *characteristic lengths* of the dissipative structures i = re, hs, sp:

$$\overline{l}_i(d) = \frac{\overline{\tilde{\mathit{æ}}}_{eff,i}}{\overline{\mathit{æ}}_{eff,i}}\, d = b_i\, d, \tag{2.37e}$$

with

$$\overline{l}_{re}(d) = 0.240\, d$$

$$\overline{l}_{hs}(d) = 0.271\, d$$

$$\overline{l}_{sp}(d) = 1.462\, d.$$

as well as the *surface area fractions* of the structures i = hs, sp:

$$\overline{a}_{hs}(d) = 1 - \overline{a}_{sp}(d) = 1 - \exp[-(d_0/d)^{a_3}]. \tag{2.37f}$$

With the physical parameters:

$$\overline{T}_{re} = 1413\,K, \quad \overline{\tilde{\mathit{æ}}}_{eff,re} = 0.380\,m^{-1}$$

$$\overline{T}_{hs} = 1329\,K, \quad \overline{\tilde{\mathit{æ}}}_{eff,hs} = 0.404\,m^{-1}$$

$$\overline{T}_{sp} = 632\,K, \quad \overline{\tilde{\mathit{æ}}}_{eff,sp} = 1.035\,m^{-1}.$$

and both of the empirical parameters $d_0 = 3.260$ m, $a_3 = 1.104$, which all result from a multiple, non-linear regression for a JP-4 pool fire, the curve $\overline{SEP}_{act}(d)$, shown in Fig. 2.18 and 2.19 is calculated.

2.6 Thermal radiation models 37

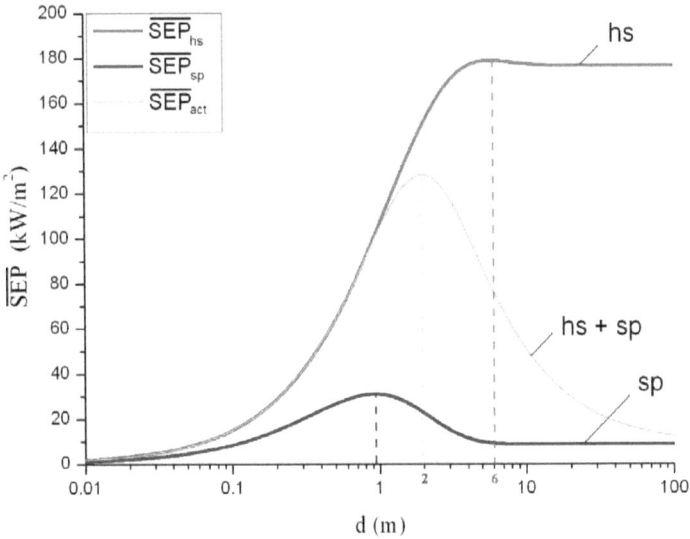

Fig. 2.18: $\overline{SEP}(d)$ curves calculated with OSRAMO II [4,7,26].

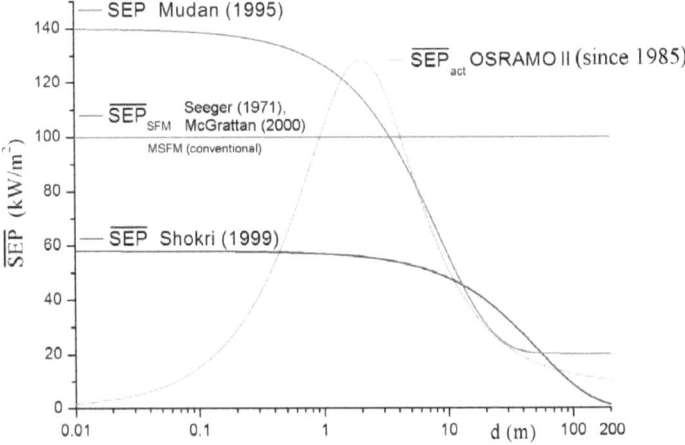

Fig. 2.19: $\overline{SEP}_{act}(d)$ curve predicted with OSRAMO II as well as according to four other radiation models [26].

With the physical parameters and the two empirical parameters from the Eqs. (2.37a-f) the curves for hot spots (hs), soot parcel (sp) as well as for the whole flame are shown in Fig. 2.18.

The discussion of $\overline{SEP}_{act}(d)$ curve, in comparison with the curves of the other mentioned models can be found in [26] (Fig. 2.19). The predicted $\overline{SEP}_{act}(d)$ curve is also in good agreement with other smoky pool fires of different fuels up to $d \approx 80$ m (Fig. 5.11, Section 5.2.5) [26].

OSRAMO III

With the stochastic radiation model OSRAMO III [7,26] from the empirically determined log-normal probability density functions for JP-4 pool fire and relating large, sooty, hydrocarbon pool fires, regarding the temperatures and temperature ranges, as well as the areas of hot spots and soot parcels, the mean surface emissive power $\overline{SEP}_{act}^{III}(d,f)$ [3,4,7,26] is calculated by:

$$\overline{SEP}_{act}^{III}(d,f) = \int_{\overline{SEP}} g_{SEP}(\overline{SEP},d,f)\, \overline{SEP}\, d\overline{SEP}, \qquad \text{for } d \geq 1 \text{ m} \tag{2.38a}$$

or:

$$\approx \int_{41.5}^{430} g_{SEP}(\overline{SEP}_{hs},d,f)\, \overline{SEP}_{hs}\, d\overline{SEP}_{hs} + \int_{6}^{41.5} g_{SEP}(\overline{SEP}_{sp},d,f)\, \overline{SEP}_{sp}\, d\overline{SEP}_{sp} \tag{2.38b}$$

$$= \overline{a}_{hs}(d,f) < \overline{SEP}_{hs}(d,f) > + \overline{a}_{sp}(d,f) < \overline{SEP}_{sp}(d,f) > \tag{2.38c}$$

with the relationship for the area fractions:

$$\overline{a}_{hs}(d,f) = \int_{41.5}^{430} g_{SEP}(\overline{SEP}_{hs},d,f)\, d\overline{SEP}_{hs} \tag{2.38c}$$

and

$$\overline{a}_{sp}(d,f) = \int_{6}^{41.5} g_{SEP}(\overline{SEP}_{sp},d,f)\, d\overline{SEP}_{sp}. \tag{2.38d}$$

With OSRAMO III the temperature and SEP regions of hot spots (hs) and soot parcels (sp) may be determined from the (previously not predictable) log-normal probability density function $\overline{g}_T(\overline{T},d,f)$, (Fig. 2.20) regarding to flame temperatures $\overline{T}$ and $\overline{g}_{SEP}(\overline{SEP},d,f)$, (Fig. 2.21) with regard to $\overline{SEP}$ for example for smoky pool fires:

$$873 \text{ K} \lesssim \overline{T}_{hs}(f) \lesssim 1653 \text{ K} \tag{2.39a}$$

$$573 \text{ K} \lesssim \overline{T}_{sp}(f) \lesssim 973 \text{ K} \tag{2.39b}$$

$$33 \text{ kW/m}^2 \lesssim \overline{SEP}_{hs}(f) \lesssim 430 \text{ kW/m}^2 \tag{2.39c}$$

$$6 \text{ kW/m}^2 \lesssim \overline{SEP}_{sp}(f) \lesssim 50 \text{ kW/m}^2. \tag{2.39d}$$

The probability density function $\overline{g}_T$ and $\overline{g}_{SEP}$ (Eqs. (2.39a-b)) for JP-4 pool fire (d = 16 m) are shown in the Figs. 2.20 and 2.21.

For the example of a gasoline pool fires (d = 25 m) the critical thermal distances (consequence or precautionary distances) are $\Delta \overline{y}_{cr}/d \approx 2.7$ (consequence model to date with $\overline{SEP}_{act}^{II} = 31 \text{ kW/m}^2$) or $\Delta \overline{y}_{cr}/d \approx 5.6$ (*new* consequence model with $\overline{SEP}_{cl}^{II}$) = 180 kW/m² or $\Delta \overline{y}_{cr}/d \approx 11.6$ (for multiple tank fires). These critical distances are in some cases much larger than the standard distances of the technical regulations TRbF.

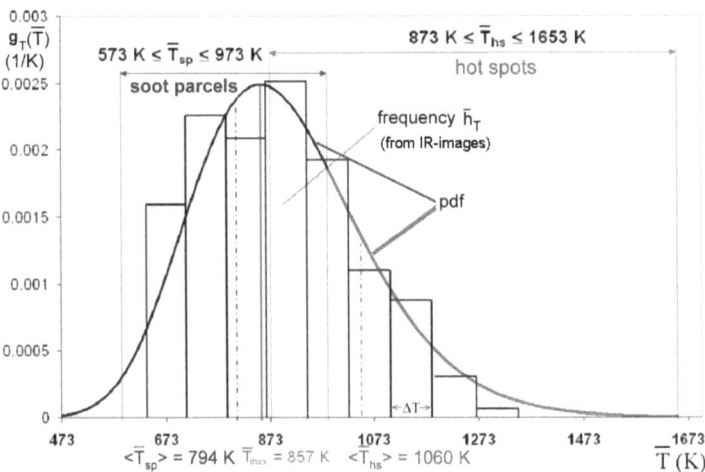

Fig. 2.20: Histogram $\overline{h}_T$ and log-normal pdf $g_T(\overline{T})$ of JP-4 pool fire (d = 16 m).

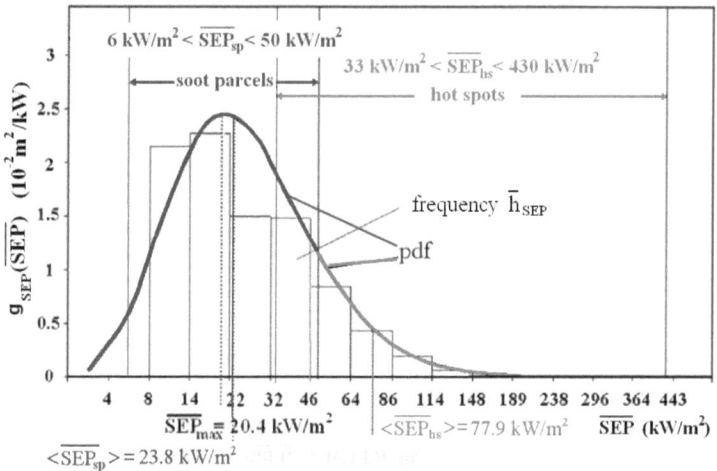

Fig. 2.21: Histogram $\bar{h}_{SEP}$ and log - normal pdf $g_{SEP}(\overline{SEP})$ of JP-4 pool fire (d = 16 m).

In the case of a probabilistic approach, the critical distance range is $0.7 < \Delta \bar{y}_{cr}/d < 3.4$ (with $\bar{g}_{SEP_{act}}(\overline{SEP})$) of a fixed threshold value (point value) of $\Delta \bar{y}_{cr}/d \approx 2.7$. The installation may be sited within this distance range [26].

2.6.3 Radiation model according to Fay

The semi-empirical radiation model according to Fay takes account of the axial dependence of $\overline{SEP}(x)$ over the whole visible flame height, so that for the *local* $\overline{SEP}(x)$ the following is valid [22]:

$$\frac{\overline{SEP}(x)}{\gamma \sigma \bar{T}_{max}^4} = k_2 x \, e^{-k_2 x}. \qquad (2.40a)$$

From Eq. (2.40a) it follows that for $x = 1/k_2$ a maximum local $\overline{SEP}_{max}(x)$:

$$\frac{\overline{SEP}(x=1/k_2)}{\gamma \sigma \bar{T}_{max}^4} = 1/e \approx 0.368 \qquad (2.40b)$$

with the scaled absorptions coefficient $k_2 = 0.0233$ m^{-1} and $\gamma\sigma\overline{T}_{max}^4 = 563$ kW/m^2 based on the example of the LNG test pool fire (d = 35 m). According to Eq. (2.40b) for a LNG pool fire a maximum $\overline{SEP}_{max} = 207$ kW/m^2 should occur at the height x = $1/k_2 \approx 42.9$ m respectively at $\overline{H} / d = 1.23$ [22].

2.6.4 Radiation model according to Ray

The semi-empirical radiation model according to Ray takes account of the *localised* probability $p(x / \overline{H})$ for the occurrence of a maximum SEP value, so that for the local $\overline{SEP}(x)$ the following is true [23]:

$$\overline{SEP}_{act}(x) = \overline{SEP}_{cl} \quad \text{for } 0 \le x/\overline{H} \le \overline{H}_{cl}/\overline{H}, \tag{2.41a}$$

and for: $\overline{H}_{cl} / \overline{H} \le x / \overline{H} \le 1$ (2.41b)

$$\overline{SEP}_{act}(x) = p(x/\overline{H})\,\overline{SEP}_{cl} + (1 - p(x/\overline{H}))\,\overline{SEP}_{s,eff}. \tag{2.41c}$$

For $\overline{SEP}_{cl}$ and $\overline{SEP}_{s,eff}$ there are the following relationships:

$$\overline{SEP}_{cl}(d) = \overline{SEP}_B (1 - e^{-d/d_{opt}}) = \overline{SEP}_B\, \varepsilon_F, \tag{2.41d}$$

with $\overline{SEP}_B = 325$ kW/m^2, $d_{opt} = 13.81$ m (for LNG pool fire, d = 13 m) (2.41e)

and $\overline{SEP}_{s,eff} = \overline{\tau}_s\,\overline{SEP}_{cl}$, with the transmissivity of smoke (2.41f)

$$\tau_s(d) = e^{-k_s \overline{c}_s L_b(d)}, \tag{2.41g}$$

as well as the mean path length:

$$L_b = 0.63\, d. \tag{2.41h}$$

2.7 Irradiance

A homogeneous, isotropic and adiabatic pool with the actual surface emissive power $\overline{SEP}_{act}(d)$ at the flame surface $\overline{A}_F$ produces, with the view factor $\varphi_{E,F}$ at any (receptor) surface element in its surroundings, at a horizontal distance Δy from the pool rim, a mean irradiance of $\overline{E}(\Delta y / d)$ [7,27]:

$$\overline{E}(\Delta y/d, d) = \tau_a \alpha_E \varphi_{E,F}(\Delta y/d)\,\overline{SEP}_{act}(d). \tag{2.42a}$$

The calculation of the view factor $\varphi_{E,F}$ is carried out according to the fundamental relation [7]:

$$\varphi_{E,F}(\overline{A}_F, \Delta y/d, \beta_F, \beta_E) = \frac{1}{\pi \Delta \overline{A}_E} \int_{A_F} \int_{A_E} \frac{\cos\beta_F \cos\beta_E}{d^2(\Delta y/d)^2} d\overline{A}_F d\overline{A}_E \, . \qquad (2.42b)$$

A simplified calculation of the view factor $\varphi_{E,F}$ may be carried out, if the flame surface assumed as a vertical circular cylinder is replaced by a corresponding quadratic area [7].

The assumption of an inclined elliptical cylinder mantle surface, which describes a real flame surface better than a circular cylinder surface, leads to a more complicated calculation of $\varphi_{E,F}$ [7].

For an approximate consideration of the flame surface $\overline{A}_F$ as a vertical or inclined circular cylinder (Fig. 2.22) the relationship for horizontal and vertical view factors $\varphi_{E,F,h}$ and $\varphi_{E,F,v}$ may be given through the calculation of the double integral Eq. (2.42b) [7], from which a maximum view factor is determined according to

$$\varphi_{E,F,h} = \frac{1}{\pi}\left[\arctan\sqrt{\frac{b+1}{b-1}} - \left(\frac{b^2-1+a^2}{\sqrt{AB}}\right)\arctan\sqrt{\frac{(b-1)A}{(b+1)B}} \right] \qquad (2.42c)$$

$$\varphi_{E,F,v} = \frac{1}{\pi}\left[\frac{1}{b}\arctan\frac{a}{\sqrt{b^2-1}} + \frac{a(A-2b)}{b\sqrt{AB}}\arctan\sqrt{\frac{(b-1)A}{(b+1)B}} - \frac{a}{b}\arctan\sqrt{\frac{b-1}{b+1}} \right] \qquad (2.42d)$$

$$\varphi_{E,F,max} = \sqrt{\varphi_{E,F,h}^2 + \varphi_{E,F,v}^2} \, . \qquad (2.42e)$$

The Eq. (2.42c-e) and corresponding relations apply to the flame shape (under the wind influence) as a cylinder flame surface and tank fire, and are shown in [25,75]. The calculation of fire with view factor accounting for the partial smoke blockage effect is also done [76]. A simplified calculation of the view factor φ can be carried out if the cylindrical surface of the flame will be replaced with a rectangular area [77]. The adoption of an inclined elliptical cylinder to the real contours of flames more realistic than a circle and cylinder which describes this complicated calculation is discussed in [72].

The relationship Eq. (2.42a) is also important to determine the $\overline{SEP}_{act}$ size, if the measurements of irradiance $\overline{E}(\Delta y / d)$ by radiometer are available, accounting that the view factor $\varphi_{E,F}$ and, consequently $\overline{SEP}_{act}$ are dependent on A_F and H [7].

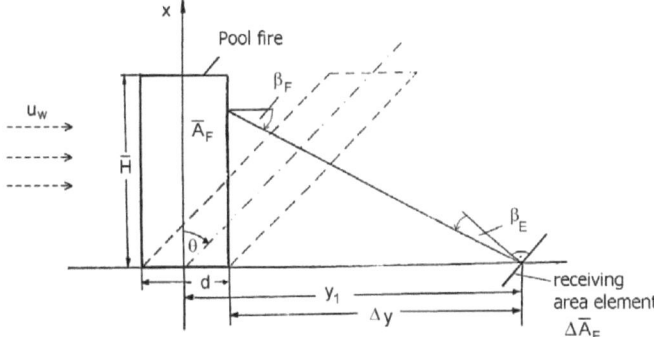

Fig. 2.22: Approximate consideration of the flame surface $\overline{A}_F$ for calculation of irradiance [7].

The distance dependency of the view factor $\varphi(\Delta y/d)$ between a circular cylinder as flame surface with diameter d and height $H = 1.7$ d, and the object to be protected is shown on Fig. 2.22.

The degree of absorption α_E of the surface of the object which should be protected as well as the atmospheric transmissivity τ_a between the flame and the object which should be protected can also only be estimated with large uncertainties. It is assumed that $\alpha_E = 1$, $\tau_a = 1$.

The calculation results obtained with the *deterministic* approach for (critical) thermal distances are valid for a pool fire and a tank fire if a critical irradiance is assumed to be a *fixed* threshold value (point value). As a fixed separation distance or a fixed effect distance or radius (in each case point values) $\Delta y/d$ results, with a threshold value $\overline{E}_{cr} = 1.6$ kW/m² for harmful effects from the time averaged irradiance $\overline{E}(\Delta y / d)$ according to:

$$\overline{E}(\Delta y / d, d) = \overline{E}_{cr}, \qquad (2.43)$$

dependent on the horizontal relative distance $\Delta y/d$ from the pool or tank rim, the value $\Delta y/d = 5.4$ (MSFM model), $\Delta y/d = 5.6$ (OSRAMO II).

The deterministic distances predicted with OSRAMO II ($\overline{E}^{II}_{act}(d)$ with $\overline{SEP}^{II}_{act}(d)$) (Fig. 2.19, Chap. 2.6.2) reach a maximum for a pool diameter of $d \approx 1.9$ m and with increasing d reduce rapidly. In contrast to this, the standard distances of the technical regulation TRbF are independent of d. For detrimental effects the standardized distances are too small by a factor of 4. For uncooled objects requiring protection (e.g. neighboring tanks) the standardized distances for d > 7 m are too conservative. For cooled tanks the standardized distances are proved to be markedly too conservative already for tanks d > 3 m.

Further from Fig. 2.18, Section 2.6.2 it is shown that the deterministic distances $\Delta y_{cr}/d$ predicted with OSRAMO II ($\overline{E}_{cl}^{II}$ with $\overline{SEP}_{cl}^{II}$ = 180 kW/m^2) in cases of the consideration of the hot, clear burning zone with $\overline{SEP}_{cl}^{II}$ = 180 kW/m^2 [1] are not dependent on the diameter d, but remain constant at a high level, e.g. $\Delta y_{cr}/d$ = 5.6 = const ≠ f(d). Finally it is noted that in cases of multiple fires with $\overline{SEP}_{cl}^{II}$ = 360 kW/m^2 [1] the resulting deterministic distances $\Delta y_{cr}/d$ = 11.6 is also not dependent on the diameter d [26].

2.8 Field models and integral models

The *field models* are generally stationary and are usually used to predict non reactive flows. Before the use of field models, the whole geometry of the simulation area must be determined in detail in a computer in a preprocessor or CAD program. Then the geometry will be divided in a variety of cells, their number is limited, in principle, only by the available hardware and software. The grid may contain the tens of thousands to several million cells. The actual calculation will be made for each individual cell, the balance equations for example, mass, momentum and possibly the energy basis of partial differential equations under certain conditions will be resolved. As the influence of neighboring cells on the particular cell must be taken into account, the solution is iteratively. When the simulation is finished, the results in the form of a multitude of variables such as flow velocity, temperature and pressure will be written for each cell. The existence of the data in two or three-dimensional fields gives this model its class names. Calculations with field models can be due to their mathematical complexity require a long computing time on modern computer systems. Through the detailed description of the reactive flows processes the field models are not restricted on specific issues but they are applicable in almost all fields of fluid mechanics. The quality of the results depends on the quality of the spatial discretization and the correct determination of the initial and boundary conditions, also the sub models.

Integral models represent a compromise between the semi-empirical models and field models. As for the field models, the solution of balance equations for momentum, energy, and possibly other sizes must be done. Thus, the models are applicable for a wider range of problems than the semi-empirical models. In contrast to the field models, the partial differential equations will be integrated and to ordinary differential equations reduced. The sizes can be determined from a condition known as an incremental calculation scheme. Therefore, also integral models are instructed by the realistic starting values. In the area of the flames, these models are not used, at present there exist no integral models for prediction of the impact of large accidental pool fires [7].

2.9 CFD simulation

The CFD (computational fluid dynamics) models are basically transient and based on the differential equations for conservation mass, species mass, momentum and energy as well as the numerous sub-models for turbulence, reaction mechanisms, soot formation and thermal radiation. The computational times are generally relatively large, but with the use of parallel computers (computing cluster) can be significantly shortened. The CFD modeling and simulation of fires is in a very promising development and will also be applied by the author (Chap. 4.4).

In contrast to the field models (Chapter 2.8) the CFD models can be used for unsteady calculations to simulate reactive flows. The use of CFD models needs, as well as in a case of the field models, a maximum fine the geometry discretization. In the computational grid the balance equations for mass, momentum and, if necessary, for other sizes such as are energy and species, based on partial differential equations are solved. The CFD models have the same advantages and disadvantages as the field models. Besides generating a sophisticated grid require the user knowledge about the strengths and weakness of sub-models available (e.g. turbulence models) and a realistic assessment of the initial boundary conditions. Moreover, modern computer systems needed to compute in a reasonable time are available. CFD models are applicable on almost all conceivable fluid mechanic problems [7].

The solution to the balance equations is then generated to the discrete nodes using arithmetic operations. The iterative methods have proven, in which the solution is gradually approached. The individual calculation steps are repeated, with the results of a run as the starting values for the next run. This approach requires initial values defined for the first step and also the boundary conditions of the balance area.

The exact solution of the balance equations requires the dissolution of the smallest occurring length and time scales. The smallest scales are Kolmogorov [78] proportional to 3/4. For typical sizes for the kinematic viscosity of 10^{-6} m^2/s is easy to calculate that for accurate three-dimensional calculations, often more than 10^{13} grid points are needed [79,80]. If the simulation is transient, dependent on time, more than 10^4 times and computing steps per grid point is needed. These demands on the memory and computational power will, of course, modern parallel computers of the near future not fulfill. Since direct numerical simulations for the most technically relevant systems currently are not feasible, the introduction of simplifying models is needed. Its simplifications are the best possible description of certain physical or chemical processes without the direct calculation, and thus reduce the computational effort. There are now models of varying complexity and accuracy for the most important problems such as the treatment of turbulence, thermal radiation and soot.

Numerical simulations are often an ideal complement to the experiments because they offer opportunities which in the experiment do not exist or are difficult to achieve. In general statements, all sizes to calculate, such as velocities, temperatures and concentrations of all species involved in the entire territory of balance are possible. In experiments, many sizes can often, only selectively be measured, in part, there are areas in which, for example, due to geometric or for safety reasons no measurements are possible. In addition, there is the option of modeling to change the boundaries of the geometry any time. Through simulations, with the parameters and geometry variations can for example, the number of expensive experiments be reduced since in the forefront, the disadvantages of certain geometries or other parameters can be detected. In order to validate the results of simulation, experiments, however, are indispensable. In principle, the differences in the experiment and simulation values must be as low as possible. Major differences between the experimental and simulated values can have various causes and should therefore always be closely investigated. Possible sources of error of the simulation are of course the first solution algorithm in the programming, or even in defining the boundary of the simulation. But even if they may be excluded the numerical solutions include always three systematic errors, they are therefore in the mathematical sense only approximate solutions:

- Model error, defined as the difference between the current flow field and the exact solution of the mathematical model;

- Discretisation error, defined as the difference between the exact solution of the conservation equations and the exact solution of the algebraic equation system;

- Convergence error, defined as the difference between the iterative and the exact solution of the algebraic equation system.

Besides the simulations exists also the measurement error, even if the error is often very small. In addition, often a physical size is not directly measured, but indirectly through several steps must be. It can happen that, for example, one or more sizes are measured directly and the requested size will be calculated. In such a case, the inaccuracies of each step and its influence on the resulting size can be estimated. In a study of the uncertainties in determining the radiant heat of a fire in a closed space, for example, in [81] uncertainties were 7% to partially over 40% depending on the experimental conditions. Given uncertainties of the measurement are not always applied, so it is often difficult to compare experiment and simulation. For comparison, the results must be calculated in the sizes useful for evaluation or can be visualized. The basis for most accurate results, however, is a suitable grid, and a precise and stable solver. Now-days, for the various steps of simulation, such as the grid generation, the convergence review and evaluation according to the calculations, a variety of software tools such as CFX, Fluent or StarCD are available.

In this work, the software packages the company ANSYS CFX and ICEM CFD grid generation software and ANSYS FLUENT and GAMBIT grid generation software are used. The programs are based on an Eulerian view of the fluid. This means that the macroscopic change of state variables in a fixed location or in a fixed control volume is registered [82]. In all generated control volumes by discretization defined calculations are carried out in order to balance the different sizes. The assumptions and simplifications of the various models and the resulting calculation steps are given in Chapter 4.

2.10 Wind influence

The counter-rotating vortices that appeared in the wind speed starting from $u_w \approx 1$ m/s follow the stream-wise direction and are close to the ground, tending to sweep fluid under and into the flame volume. It was also observed [9] that the rotational structures formed without the presence of a cross-wind, while coherent, are continuously moving and thereby produce no time-mean effect at a fixed location. In this sense they have no time-mean definition. While the rotation of the structure may be steady in temporal sense the position of the structure is not because it is being continuously advected up the fire plume. The vertical structures that are formed in no wind conditions are still formed on the upwind (or winward) side of the fire. In addition, new rotational structures are formed on the downwind (or leeward) side of the fire and appear to have a relatively steady mean. On the upstream side of the fire, vertical structures are observed [9] that appear to have characteristics very similar to those formed in the absence of a crosswind. The rotational structures increase in a scale from the toe of the fire. From the video sequences [9], the upwind structures appear periodic because the structures pass a given point in space with a relatively constant frequency. In a case of a cross-wind the structures are quickly advected downstream away from the toe before growing to large length scales [9].

The vertical structures formed on the downwind side of the fire are distinct from those formed on the upwind side of the fire. On the downwind side, the dominant direction of rotation is axial. The columnar rotational structures formed on the downwind side of the fire have a time-mean definition. The dominant direction of the smoke and flame streaks is upward. The streaks are not completely vertical but sweep inward from each side. For the time period of photographs according to [9] the structures have a time-mean rotational velocity. The plume is divided into two counter-rotating plumes. Each columnar vortex grows in the size but its base moves farther downstream from the fire and then blows out. Under the wind influence the location of columnar vortices is on the lee side of the pool or downwind side of the fire depending on the strength of the crosswind. The higher the wind velocity is, the larger both the tilt of the fire plume and the ground surface area covered by the fire are (Fig. 2.23). Under a certain wind and terrain conditions, the entire fire can be spun into a single columnar vortex. This condition is termed a "fire-whirl" and produces heat fluxes that are particularly destructive [9]. In the calm conditions a wrinkled surface is being rolled into larger and large structures as the height above the fire increases. For a larger fire, it is difficult to visualize the flame sheets away from the toe of the fire because they are hidden behind the smoke layer. In a crosswind, the large columnar vortices are relatively clear of smoke [9].

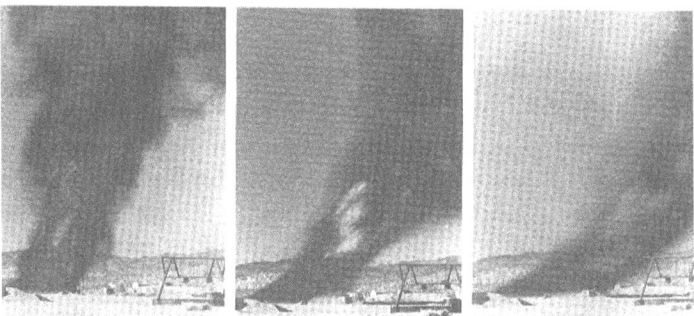

Fig. 2.23: JP-8 pool fire (d = 20 m) under the wind influence (with increasing wind velocity from left to right) [9].

The effect of wind on thermal radiation from the free-burning aviation fires (0.9 m ≤ d ≤ 2.4 m) has been studied experimentally in wind tunnel tests [11]. Wind causes the flame to tilt and to blow the smoke away from the flame [11].

The wind may qualitatively affect the rate of baroclinic vorticity production because the air entrained into the fire is not quiescent, but has a momentum. The entrainment of air with a mean momentum results in higher advection of the vortices away from the toe of the fire. Wind produces baroclinic vorticity in a fire signed as a pair of counter-rotating vortices. The columnar vortices are swept downstream from the pool. As the vortex move away from the leeward side of the fire, fuel vapor can not longer be entrained so at a certain distance from the pool vortices start to quench. The fuel/air eddies are convected downstream from the pool due to the horizontal air entrainment. Evidence for combustion near the ground level downstream from the pool can be seen in soot footprints. The phenomenon is called "flame drag" [9,11,16]. In the experiments [11] the horizontal radiation profile of free burning fire of AVGAS and JP-4 it has been measured by water cold radiometers with 140° view angle [11]. The radiations were measured in three directions: passing the center of the fuel pan, parallel (upwind and downwind), and perpendicular (crosswind) to the direction of wind [11]. The same measurements were done in a still air. In the downwind direction the flame is elongated by the wind with smoke moving ahead of the flame. Smoke blockage occurred at x = 1 m. To prevent that, the radiometer nearest to the fire was placed at y/d = 3 and x = 0.46 m was used for all radiometers [11]. It is found that the SEP and E are affected by the wind and this effect diminishes as the distance increases. Thermal radiation downwind of the pool increases with wind speed, but reaches a maximum between 6.7 m/s and 8.9 m/s and decreases at higher speeds. The large value of SEP and the SEP and downwind is due to the flame bending. The decrease is due to the excessive flame bending, which makes the flame appear smaller to the radiometer. The same behavior is seen for the SEP and E in a crosswind direction but their variations is smaller than in a case of SEP and E in downwind direction. In the experiments the difference in the radiation of light sooty AVGAS and high sooty JP-4 it was not observed due to the reason that the wind blow the smoke away from the fire so the radiometers were not covered by the smoke [11]. Wind causes significant radiation increases in the downwind direction of a fire, but decreases the data scatter compared with no wind data [11]. The magnitude of this increase at various wind speeds is relatively small, however. Wind affects the radiation level slightly in the upwind and cross-wind directions [11].

3 Experiments

The flame emission temperatures and Surface Emissive Power (SEP) were measured by IR thermography system. For the measurements of irradiance wide angle radiometers were positioned depending on the expected heat radiation in certain intervals of the flame. The flame length was with the photo and video recordings determined.

3.1 Pools

To study the dynamics of large pool of flames and the determination of various properties such as flame length and temperature and emitted thermal radiation the large scale experiments have been done on pool fires with different fuels and circular pool diameters of 2 m, 8 m, 16 m and 25 m. The tests are done at the research area of German research center for air and space travel (Deutschen Forschungs- und Versuchsanstalt für Luft- und Raumfahrt - DLR) in Trauen near Munster (Fig. 3.1) [3,4,30].

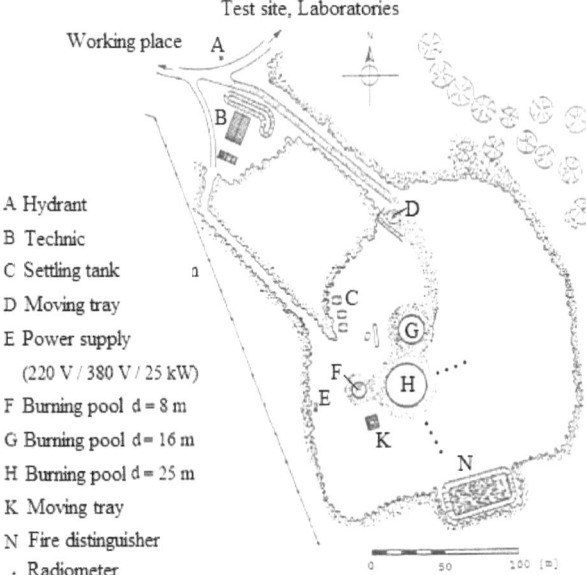

A Hydrant
B Technic
C Settling tank
D Moving tray
E Power supply
   (220 V / 380 V / 25 kW)
F Burning pool d = 8 m
G Burning pool d = 16 m
H Burning pool d = 25 m
K Moving tray
N Fire distinguisher
· Radiometer

Fig. 3.1. DLR test site in Trauen [3,30].

52 3 Experiments

The experiments have been carried out in the forest area with a side length of about 150 m. The pools are made of concrete poured, and are equipped with blue basalt stones. In addition, the soil under the pool is isolated with an oil film to prevent the leaking of fuels. The pools were up to about 10 cm below the ambient levels filled with water and the fuel was stratified above it (Fig. 3.2) [3].

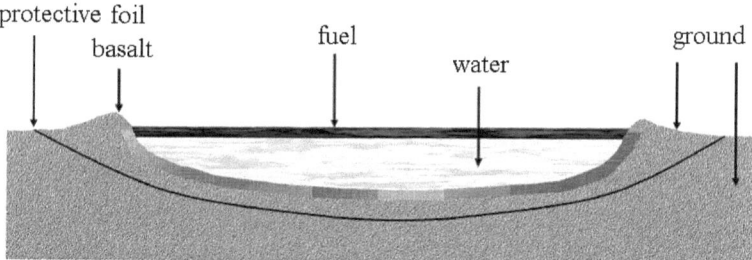

Fig. 3.2: Construction of pool according to [30].

3.2 Fuels

There were various fuels used for this work: premium gasoline, normal gasoline, pentane and JP-4 (Jet fuel 4, aircraft fuel) as a mixture of 80% kerosene and 20% normal gasoline.

3.3 IR thermographic camera system

The measurement of emission temperatures and surface emissive power (SEP) of flame was done with an infrared IR thermography system of delivers ThermaCAM^{TM} Researcher by the manufacturer FLIR Systems GmbH (and by Hughes Aircraft Company [3]). The system consists of an infrared transducer, a black-and-white video camera, a digital processor with a color screen and a video-mixing unit. The IR system has a 4-bit resolution, the temperature division into 16 areas at a maximum sensitivity of $\Delta T = 0.1$ K [3] (The thermal sensitivity of the device is in the area of T = 0.08 K at 30° C with standard 50 Hz and the accuracy of reading is ± 2% [3]). The frequency of the image recording system is 20 frames per second corresponding to a time resolution of $\Delta t = 0.05$ s, a maximum temperature of T = 0.1 K resolution and a spatial angle of $\Omega \approx 0.04623$ sr [3,4]. Detected wavelength range is 2 μm < λ < 5.6 μm. The system has a maximal image resolution of 640 * 480 pixels in color [3]. (The detectable wavelength ranges from 7.5 microns to 13 microns [3]). The distance between the IR thermography system and the flame was 50 m. The emission level s is set to $\varepsilon_F = 0.98$.

The measured flame temperatures are used for determination of surface emissive power (SEP) by Stephan Boltzman Law.

The measured temperature distribution structures assign visible flames. More details on IR thermography system and results of measurements are described in [3].

3.4 VIS camera system

The measurement of flame lengths is made with two S-VHS video cameras and digital cameras. In the visible spectral range (0.4 µm < λ < 0.75 µm) the JP-4 pool fires were photographically registered. With a video-mixing unit, a superposition of video and thermographic images is possible to get the temperature distribution in visible spectrum.

For recording of the VIS-structures of the flames a 16 mm film camera of the type "Beaulieu R16" is used. The camera had changeable directions and had a 12-120 mm zoom lens type Angenieux ". The camera was mounted on a tripod and an external battery pack supplied with power. With few exceptions, the host frequency was 50 frames per second. To verify the recording frequency, a warning lamp, which illuminates a defined frequency, the test site and also position on the film is used.

By superposition IR and VIS fields of the flames the fluid dynamic structures as soot parcels and hot spots and was allocated.

3.5 Radiometer measurements

For the irradiance measurements were ellipsoidal radiometer Meditherm Company (USA) and infrared sensors of the company INC used (United States) used. This is a windowless infrared radiometer applicable for measurements of surface emissive power and temperature [3]. There were three ellipsoidal radiometers near the flame, six infrared sensors away from the flame up. Fig 3.3 schematically represents the ellipsoidal radiometer used here.

The head of the radiometer is constructed in the form of ellipsoidal cavity, with the radiometer opening in one of the focal points of ellipsoids is located. A sensitive radiation thermopile is in another burning point indicated. The maximum opening angle is 180° (and thus the space angle $\Omega \leq 2\pi$ sr), but in the experiments was only an opening angle of 95° used.

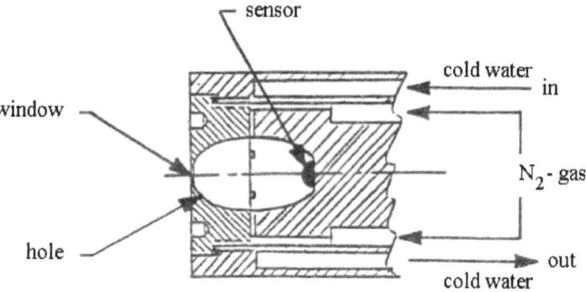

Fig. 3.3: Schematic construction of the radiometer [3].

The detectable wavelength range, depending on the window material is between 10 nm and 1 mm. During the trials the ellipsoidal radiometers were constantly cooled by using N_2 and water. For the cooling of the detectors up to 87 K (–186 °C) the cryostatic temperature regulator is used [3].

To measure the irradiance E ($\Delta y/d$, t) depending on the relative distance $\Delta y/d$ from the pool edge and time t the ellipsoidal radiometer made by the company Medtherm Coperation is used [3,4]. The radiometers had a wavelength range of 0.6 µm < λ < 15 µm (the reception wavelength area of ellipsoidal radiometer is between 0.2 microns and 7 microns [3,4]) with the absorption of the detector element of the radiometer α = 0.92.

The arrangement of the radiometer was conducted according to Fig. 3.3 in a relative distance $0 \leq \Delta y \leq 1$ from the pool edge of the various flames. The radiometer has a strength to keep clean (possibly soot which could lead to pollution and may disturb the measurements) during the experiments. Performance of the radiometers each is in the range of 200 kW/m², 100 kW/m² and 50 kW/m² [3,4].

3.6 Wind measurements

Registration for the wind conditions is done by 10 m high tower with sensor mounted at heights of 2 m, 6 m and 10 m. The measuring station registered the wind speed and direction with a time resolution of 3 s [3].

4 Some important topics of CFD used in this work

The numerical fluid mechanics, Computational Fluid Dynamics (CFD) is based on the solution of the conservation and transport equations for mass, momentum, energy and mass fractions of various species. In a case of turbulent flows, the Reynolds averaged transport equation for modeling of turbulence is solved in addition to these equations. Most commercial CFD codes use finite volume method for solving the above mentioned equations. Fig. 4.1 shows is the theoretical way to solve a fluid mechanic problems by CFD simulation e.g. contained in ANSYS CFX and FLUENT softwares.

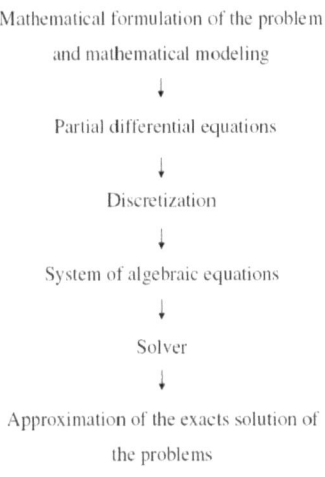

Fig. 4.1: Procedure of solving fluid mechanic problems by CFD simulation.

4.1 The conservation equations in fire modeling

Combustion processes consist of a variety of processes such as flow, chemical reaction and molecular transport, for example, heat conduction, diffusion and radiation. Such a chemically reactive flow can be described at any time and anywhere by characteristics such as pressure, density, temperature, speed and composition of the mixture [68]. These sizes vary depending on time and space. Some sizes in these chemically reactive flows have the property that they are independent of the processes. These include energy, mass and the momentum.

4 Some important topics of CFD used in this work

The mathematical modeling of transport processes is based on conservation equations.

The basic idea is that the storage of a conservation size in a control volume is equal to the sum of flows from the volume at the volume surface. Both, the convective as well as the diffusive flows are covered. Generally, overall conservation equation of value Φ can be presented as a change in the density function of a general state in Euler's coordinates:

$$\underbrace{\frac{\partial(\rho\Phi)}{\partial t}}_{\text{Accumulative}} + \underbrace{\nabla(\rho\vec{v}\Phi)}_{\text{Convective}} = \underbrace{\nabla(D_\Phi \text{grad}\Phi)}_{\text{Diffusive}} + \underbrace{S_\Phi}_{\text{Source}} \quad . \tag{4.1}$$

The symbol ∇ stands for the nabla operator. In this work the partial differential equations in the three spatial directions indicate:

$$\nabla = \left[\frac{\partial}{\partial x}, \frac{\partial}{\partial y}, \frac{\partial}{\partial z}\right]. \tag{4.2}$$

According to Eq. (4.1) the temporal change of size Φ by the convective and diffusive transport with the general exchange coefficients D_Φ is determined, as well as through local sources or sinks S_Φ. The nature of the sources may be completely different and depends on the conservation size. Source terms can be temporally and spatially variable and dependent on one or even several different conservation sizes. In species conservation equation occur, for example source terms for the change of species mass fractions as a result of chemical reactions. The simulation of pool flames needed conservation equations of a reactive gas-soot mixture in the gravity, obtained by substitution of variable Φ by the following variables:

- mass, $\Phi = 1$ for the continuity equation,
- speed, $\Phi = u$ for the momentum conservation equation,
- energy, $\Phi = E$ for the energy conservation equation,
- species, $\Phi = n_i$ for the species conservation equations.

In the following chapter the conservation equations of mass, species, momentum and energy for a complete description of the interaction between the combustion and the flow field are presented.

4.1.1 Overall mass conservation

An essential component of a system is the mass. The mass has no sources inside a finite volume and in contrast to the molecular species no transportation. For mass density based on the overall density it follows:

$$\frac{\partial \rho}{\partial t} = -\nabla(\rho u). \tag{4.3}$$

Eq. (4.1) is also known as continuity equation and is designated for compressible flows. In the special case of incompressible flows or density dependent flows the continuity equation is simplified to:

$$-\nabla = 0. \tag{4.4}$$

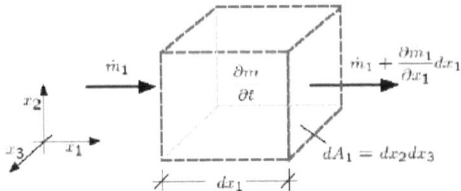

Fig. 4.2a: Mass derivation balance in a differential volume element $dV = dx_1 dx_2 dx_3$.

In Fig. 4.2a a differential volume element $dV = dx_1 dx_2 dx_3$ according to the Eulerian consideration is shown. The infinitesimal mass dm in the element is given by $dm = \rho dV = \rho dx_1 dx_2 dx_3$. According to the mass balance, a temporal change of the mass in flow element in the volume element dV is equal to the difference between a mass inflow and outflow. In the x_1 direction through the area $dA_1 = dx_2 dx_3$ a mass inflow is $\dot{m}_1$ and an mass outflow is $\dot{m}_1 + (\partial \dot{m}_1 / \partial x_1) \partial x_1$ with $\dot{m}_1 = \rho v_1 dA = \rho v_1 dx_2 dx_3$. The product ρv_1 is mass flow through the unit surface perpendicular to the x_1 coordinate and gives mass flow density. The difference between incoming and exiting mass flow in x_1 direction is:

$$\dot{m}_1 - \left(\dot{m}_1 + \frac{\partial (\rho v_1)}{\partial x_1} \partial x_1 \partial A_1 \right) = -\frac{\partial (\rho v_1)}{\partial x_1} \partial x_1 \partial x_2 \partial x_3. \tag{4.5}$$

The difference between the incoming and exiting mass flows in all three coordinates and directions divided by $dx_1 dx_2 dx_3$, gives the mass balance in volume element dV written as:

$$\frac{\partial \rho}{\partial t} = -\left(\frac{\partial(\rho\vec{v}_1)}{\partial x_1} + \frac{\partial(\rho\vec{v}_2)}{\partial x_2} + \frac{\partial(\rho\vec{v}_3)}{\partial x_3}\right). \tag{4.6}$$

The Eq. (4.6) can be also written as:

$$\frac{\partial \rho}{\partial t} = -\nabla(\rho\vec{v}). \tag{4.7}$$

This differential mass conservation equation is included in numerical solution of the flow as continuity equation [83,84]. Since that contains only momentum and convective term, the mass balance equation is mathematically easier to solve than the other balance equations.

4.1.2 Species mass conservation

The mass balance equation is based on the total mass of a system. For a system with more substances in which the distribution of various species of interest is also connected with chemical reactions and separately accounting of each species is required. Looking at the mass of various species, the partial density ρ_i of the ith component is given by $\rho v_{1,\alpha}$ where $\alpha = 1,2,3$. When the speed of the ith component in a multi-substance system (or mixture) is defined as the $v_{1,\alpha}$, with $\alpha = 1,2,3$, the summation of all components gives a flux density:

$$\rho\vec{v} = \sum_{i=1}^{n} \rho_i v_{i\alpha}, \alpha = 1,2,3. \tag{4.8}$$

The temporal change in the partial mass per unit volume resulted from a partial mass flow over the border area as well as and the balance from the chemical reactions in the volume element have formed partial mass. Conservation equations for the partial mass will be written analogous to the continuity equation for components:

$$\frac{\partial \rho}{\partial t} = \frac{\partial(\rho_i v_{i\alpha})}{\partial x_1} + \dot{m}_i. \tag{4.9}$$

The speed of the ith component consists of the main velocity, $v_{i,\alpha}$, $\alpha = 1,2,3$ and the diffusion speed ($j_{i,\alpha} / \rho_i$), $\alpha = 1,2,3$:

$$v_{i\alpha} = v_\alpha + \frac{j_{i\alpha}}{\rho_i}, \alpha = 1,2,3. \tag{4.10}$$

It follows the balance equation for ρ_i:

4.1 The conservation equations in fire modeling

$$\frac{\partial \rho}{\partial t} = \frac{\partial(\rho_i v_\alpha + j_{i\alpha})}{\partial x_\alpha} + \dot{m}_i \quad \alpha = 1,2,3. \tag{4.11}$$

For example in the software CFX and FLUENT, for accounting of mass fraction $Y_i = \rho_i/\rho$, the following equation is given:

$$\frac{\partial(\rho Y_i)}{\partial t} = -\nabla \rho Y_i u + \nabla(D_i \nabla \rho Y_i) + S_i. \tag{4.12}$$

Based on relationship in Eq. (4.12) for all species i and subsequent summation the global mass balance equation logically follows that the summation of the source terms of chemical reactions and the diffusion terms is equal zero. Since all mass Y_i breaks to one sum and the global mass conservation is resolved in any case, the conservation Eq. (4.12) has to be resolved only for i–1 species. The mass fraction of species can be easily found by difference. Due to relatively small number of balance equations, the replacement of a partial differential equation by an algebraic relation makes a substantial reduction of the time.

In the chemical balances, for example dissociation balance, however, the balance of species must be described by the forward and backward kinetics reaction. Such reactions have great speed constants, so that the equations for each species are strongly linked. There is therefore a rigid system of equations whose solution requires specific mathematical procedures. For such systems the rate constants are not usually available in the literature and can only be roughly estimated.

4.1.3 Momentum conservation

The temporal change of the momentum density ρv_α, $\alpha = 1,2,3$ can be done by external forces to the surface of the volume element dV, and by so-called far distance forces or volume forces to the mass.

The momentum flux density is composed of the convective part $v_\alpha(\partial v_\beta)$, $\alpha,\beta = 1,2,3$ and the part of power density tensor $P_{\alpha,\beta}$, $= 1,2,3$ as shown in Fig. 4.2b. The convective part of the momentum density in the volume can be written analogous to the mass conservation:

$$\frac{\partial(\rho v_\beta)}{\partial t}\bigg|_{conv.} = -\frac{\partial(\rho v_\alpha v_\beta)}{\partial x_\alpha}. \tag{4.13}$$

The momentum flux density in the volume element dV by surface forces can be calculated as:

$$\frac{\partial(\rho v_\beta)}{\partial t}\bigg|_{surf.} = \frac{1}{dV}\left(P_{\alpha\beta}dA - \left(P_{\alpha\beta}dA_\alpha + \frac{\partial P_{\alpha\beta}}{\partial x_\alpha}dx_\alpha dA_\alpha\right)\right) = -\frac{\partial P_{\alpha\beta}}{\partial x_\alpha}, \quad \beta = 1,2,3. \quad (4.14)$$

Here is only the gravitational force g_β, $\beta = 1,2,3$, as a mass force for combustion processes considered. For the mass force in a volume it follows: ρg_β, $\beta = 1,2,3$.

For the momentum equation based on the three above mentioned parts it follows:

$$\frac{\partial(\rho v_\beta)}{\partial t} = \frac{1}{\partial x_\alpha}(\rho v_\alpha v_\beta + P_{\alpha\beta}) + \rho g_\beta, \quad \beta = 1,2,3. \quad (4.15)$$

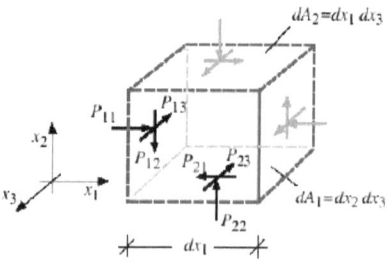

Fig 4.2b: Derivation of momentum balance at a differential volume element dV.

4.1.4 Energy conservation

There are many kinds of energy in a system which can be converted into each other. The description of the energy requires a higher coupling of the mass and momentum. In this chapter, the conservation of energy is described by a separate view of kinetic, potential and the total energy [85].

For example, while in solids, the heat conduction is only process, in fluids with relatively low temperature the heat transfer by convection is the dominant process. In reactive flows, such as pool flames, presented in this work, beside the convection, energy source terms due to chemical reactions and the energy transport due to the heat radiation are important. An energy conservation equation presented in CFX and FLUENT software is:

$$\underbrace{\frac{\partial(\rho h_{tot})}{\partial t}}_{} = \underbrace{\frac{\partial p}{\partial t}}_{\text{Work of pressure forces}} \underbrace{- \nabla(\rho u h_{tot})}_{\text{Convective transport}} + \underbrace{\nabla(\lambda \nabla T)}_{\text{Heat conduction}} + \underbrace{\nabla\left(\mu \nabla u + (\nabla u)^T - \frac{2}{3}\nabla u \delta u\right)}_{\text{Viscous work}} + \underbrace{S_h}_{\text{Source term}}$$

(4.16)

where source term can contain chemical reaction or radiation term. Total enthalpy of the system consists of static and kinetic enthalpy:

$$h_{tot} = h_{stat} + \frac{1}{2}u^2.$$ (4.17)

The specific static (thermodynamic) energy of the system consists of internal energy U and the state of the fluid:

$$h_{stat} = U + \frac{p_{stat}}{\rho_{stat}}.$$ (4.18)

The temporal change of the energy (Eq. (4.16)) is due to the work of pressure forces, the convective transport, the heat conduction and the viscous work. In addition, the source term (S_h) can include chemical reactions or radiation. For the energy conservation movement, the total energy source must therefore be free. So, the source term describes only the transformation of individual forms of energy into each other, such as the change of inner energy through the heat of chemical reaction process or the conversion of potential into kinetic energy through work. That is included in the sub-models described in the Chapter 4.2.

4.2 Sub-models in fire modeling

Mathematical modeling is the image of the resulting physical and chemical processes caused by states or state changes in a system, described with the help of mathematical relationships. Often, these equations due to their complexity can not be solved analytically, so the numerical solution methods are needed. The first step is a spatial discretization of grid structures in finite elements or subdivided volumes. The solution of the balance equations can be done at the discrete nodes using arithmetic operations. They have iterative methods, in which the solution is gradually approximated. The individual steps are repeated, the results of a run as the starting values for the next step are used. This approach requires defined start values for the first step and also defines the boundary conditions of the investigated area.

The exact solution of the conservation equations requires resolution of the smallest occurring length and time scales. The smallest scale is Kolmogorov [78]. For the special case of fully developed isotropic turbulence, Kolmogorov derived that the relation: $e(k) \propto k^{-5/3}$ is valid for completely developed turbulence [68]. To resolve the smallest scale a Direct Numeric Simulation (DNS) should be used. For a typical turbulent flow with $R_l = 500$, a $l_0/l_k = 100$, thus, about 1000 grid points in one dimension are needed, 10^9 grid points in three dimension to resolve the smallest turbulence eddies. The Navier-Stokes equation is time dependent so at least 1000 time steps are needed to mimic a turbulent combustion process, the number of computational operations exceeds 10^{14} [68].

The demands on the memory and computing power, even by use of modern parallel computers of the near future can not be fulfilled. So, direct numerical simulations for the most technically relevant systems are currently not yet feasible and introduction of simplified models are needed. The simplifications contained therein can give a possible good description of certain physical or chemical processes without their direct calculation and thus reduce the computational effort. Different models with various complexity and accuracy for the most important problems such as the treatment of turbulence, heat radiation and soot formation have been developed. Numerical simulations are often an ideal support to the experiments due to the opportunities to calculate situations which can not be obtained experimentally or are difficult to obtain. In experiments, in contrary to CFD the many sizes are often only selectively measured, in part, there are areas where for example measurements are not possible due to the geometric or the safety reasons. CFD offer many of possibilities, e.g. in modified various boundary or geometries and has ability to calculate all sizes such as velocities, temperatures and concentrations of all the species in the entire recorded territory.

In this work, the software packages ANSYS CFX and mesh generator ICEM CFD are used as well as the ANSYS FLUENT and its mesh generator GAMBIT. The programs are based on Euler's view of the fluid. This means that the macroscopic change of state variables is registered at a fixed location or in a fixed control volume [82]. In contrast, the microscopic Lagrangian representation of the movement of a particle through a system is prosecuted. Basically, a combination of the two possible methods is possible, such as the simulation of the combustion of coal particles in fire or burners where the coal particles are with the Lagrange method pursued and the flame gases with the Euler method described. In all of the control volumes produced by discretization, partial calculations are carried out and resulted calculation steps are used to hold with balance sizes. The assumptions and simplifications of the individual models and the resulting calculation steps are described in subsequent chapters.

4.2.1 Modeling of turbulence

In principle, a flow can be characterized as laminar or turbulent, with a smooth transition between two states. Laminar flows are characterized by parallel flow lines in the main flow direction. Disorders are relatively quickly restored and the initial flow conditions recovered. In contrast, in turbulent flows occur eddies, which by disturbances such as shear walls between impulse and fluid or fluid exchange between layers rise with different speeds. The vortex effect increases transportation of all sizes across the balance sheet or contrary to the main flow direction. The flow through the eddy formation is irregular, chaotic and difficult to predict. The transition from laminar to turbulent flow occurs at a characteristic Reynolds number

$$\text{Re} = \frac{\rho u l}{\mu} = \frac{u l}{\nu} \qquad (4.19)$$

with the density ρ and a dynamic viscosity μ or a kinematics viscosity ν. Furthermore, for one system the characteristic speed u and a characteristic length l must be determined. Thus, for example, in tube flows where l is usually equal to the pipe diameter, the transition point from laminar to turbulent flow is then at Re $\approx$ 2000. In turbulent flows carried by momentum and energy exchange between the vortices a decay of larger vortices to smaller is happening. Here, the turbulence energy will transfer to internal or thermal energy. The largest length scales correspond to the geometric dimensions of the system, so-called integral length l_0. Through the steady decay a cascade of energy are created, which ends with the smallest measured, Kolmogorov length l_k. By the Kolmogorov length is the kinetic energy of small vortices so dissipated that the half time of rotation of one vortex is equal to the time for the diffusion along the diameter l_k. Thus, under the l_k the diffusion is faster than the turbulent transport and turbulent processes can no longer occur. The distribution of turbulent kinetic energy to the different length scales may be determined from the

4 Some important topics of CFD used in this work

spectral energy density $e(k_e)$. It is partly dependent on the wave number k_e, which is equal to the reciprocal of the turbulent length l. It is linked with the average specific turbulent kinetic energy per unit mass q as follows [86]:

$$q(\vec{r},t) = \int_0^\infty e(k_e,\vec{r},t)dk_e. \qquad (4.20)$$

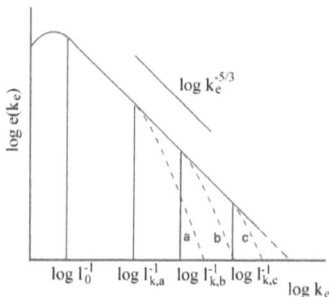

Fig. 4.3: Turbulent energy spectrum: l_0 symbolizes the integral length, $l_{k,a}$, $l_{k,b}$, $l_{k,c}$ which are the Kolmogorov-length dimensions for three Reynolds numbers $a = Re_1 < b = Re_2 < c = Re_3$ [80].

The logarithmic application in Fig. 4.3 shows that in a three-dimensional flow the spectral energy density declines at smaller length scales proportionally to $k_e^{-5/3}$. This means that the main part of turbulent kinetic energy in located in the large vortices. The influence of the Reynolds numbers on the spectral energy density is based on the three curves: a, b and c. With increasing Reynolds number from curve a to curve c moves the Kolmogorov length to smaller wavelengths, i.e. here the turbulence dominates in transport of increasing number of smaller vortices. To describe the turbulence degree a geometry dependent Reynolds number and turbulent Reynolds number can be used [86]:

$$R_l = \frac{\overline{\rho}\sqrt{2ql_0}}{\overline{\mu}}. \qquad (4.21)$$

The Kolmogorov length can be calculated by the turbulence Reynolds number:

$$l_k = \frac{l_0}{R_l^{3/4}}. \qquad (4.22)$$

4.2 Sub-models in fire modeling   67

Due to the ongoing conversion of kinetic energy in the internal or thermal energy turbulent flows require a permanent supply of power for its maintenance. This may be caused, for example, by a constant flow based on gravitational forces, in nature or in the pipeline by pumping effect. The spatial discretization of very small length scales of the smallest vortices and thus the direct numerical calculation in practical applications with today's computer technology is not possible. Therefore, the turbulence without their exact resolution use certain assumptions described below. Some of the most commonly used models for turbulence modeling will be described below.

4.2.1.1 k-ε and k-ω models

Generally, turbulence model is presented by a partial differential equation. Contains the equation new unknown, such as a turbulent dissipation $\varepsilon(x, y)$, a further modeling of equations is possible. Depending on how many partial differential equations are used there exist one-equation models, two-equation models, etc. The k-ε turbulence model is a two-equation model. Two-equation models k-ε and k-ω use the gradient diffusion hypothesis to relate the Reynolds stresses to the mean velocity gradients and the turbulent viscosity. The turbulent viscosity is modelled as the product of a turbulent velocity and turbulent length scale. Both the velocity and length scale are solved using separate transport equations (hence the term 'two-equation'). In two-equation models, the turbulence velocity scale is computed from the turbulent kinetic energy, which is provided from the solution of its transport equation. The turbulent length scale is estimated from two properties of the turbulence field, usually the turbulent kinetic energy and its dissipation rate. The dissipation rate of the turbulent kinetic energy is provided from the solution of its transport equation [83,84]. The model describes two partial differential equations for development of the turbulent kinetic energy $k = \dfrac{\overline{u'_i u'_i}}{2}$ defined as the variance of the fluctuations in velocity and the turbulence eddy dissipation ε (the rate at which the velocity fluctuations dissipate)

$$\varepsilon = \nu \overline{\left(\dfrac{\partial u'}{\partial x_k}\right)^2} \; [83,87].$$

The equation of k-ε turbulence model with kinetic energy k is given by the following relations:

$$\rho\frac{\partial k}{\partial t}+\rho\overline{u}_j\frac{\partial k}{\partial x_j}=c_\mu\rho\mu_t\left(\frac{\partial\overline{u}_i}{\partial x_j}+\frac{\partial\overline{u}_j}{\partial x_i}\right)-\rho\varepsilon+\frac{\partial}{\partial x_j}\left(\left(\mu+\frac{\mu_t}{\sigma_k}\right)\frac{\partial k}{\partial x_j}\right),\qquad(4.23)$$

$$\rho\frac{\partial\varepsilon}{\partial t}+\rho\overline{u}_j\frac{\partial\varepsilon}{\partial x_j}=c_{\varepsilon1}\frac{\varepsilon}{k}\tau_{i,j}\frac{\partial\overline{u}_i}{\partial x_j}-c_{\varepsilon2}\frac{\varepsilon^2}{k}c_\mu\rho\mu_t\left(\frac{\partial\overline{u}_i}{\partial x_j}+\frac{\partial\overline{u}_j}{\partial x_i}\right)\frac{\overline{u}_i}{\partial x_j}-c_{\varepsilon2}\rho\frac{\varepsilon^2}{k}+$$

$$\frac{\partial}{\partial x_j}\left(\left(\mu+\frac{\mu_t}{\sigma_\varepsilon}\right)\frac{\partial\varepsilon}{\partial x_j}\right).\qquad(4.24)$$

In the above equations are few, simplified model assumptions incorporated. This restricts the scope of equations. In these equations appear still unknown coefficients. These are supplemented by the consideration of simple flow fields. The parameter $c_{\varepsilon1}$ is calibrated by a homogeneous shear train at equilibrium. The size $c_{\varepsilon2}$ follows from the decay of homogeneous grid turbulence. The turbulent Prandl number σ_ε is derived from an analysis of the logarithmic area of a wall turbulent boundary layer. From the context of the anisotropic parameter c_μ the eddy viscosity can be averaged to $v_t = c_\mu$ (k^2/ε). For the turbulent boundary a value for c_μ is used. For the standard k-ε model, in the literature the following constants can be found:

$c_\mu = 0.09$, $c_{\varepsilon1} = 1.44$, $c_{\varepsilon2} = 1.92$, $\sigma_\varepsilon = 1.3$, $\sigma_k = 1$.

The standard k-ε turbulence model is characterized by its numerical stability and especially relatively small computing time and is easy to implement. Within ANSYS CFX and FLUENT, the k-ε turbulence model with buoyancy terms is available with the scalable wall-function approach added to improve robustness and accuracy when the near-wall mesh is very fine. If the full buoyancy model is being used, the buoyancy production term is modelled as:

$$P_{kb}=\frac{\mu_t}{\rho\sigma_\rho}g\nabla\rho.\qquad(4.25)$$

This buoyancy production term is included in the k and ε equation if the buoyancy turbulence option is set as an option [83,84].

4.2.1.2 Large Eddy Simulations

Large Eddy Simulation (LES) presents a completely different approach in calculating turbulence as the Reynolds averaged models and allows a much more accurate description of a turbulent flow field. Turbulent flows include a variety of eddies, whose time and length scales cover a wide area. The variation of length scales is given in Fig. 4.4a. Large eddies have a significantly higher energy and by their size and strength they essentially determine the transport of the balance sheet sizes normal to a main flow direction.

In Large Eddy Simulations are large eddy directly calculated and small energy eddies are modeled. It is due to user, to discretize the computing field so that the cell dimensions are significantly smaller than the length scale of the energy rich vortices. The vortices for which the computing grid is not enough resolved are calculated by sub grid scale models (SGS models). A well known SGS model is e.g. Smagorinsky [87,88], based on an Eddy viscosity approach similar to the RANS models. In addition to the separate calculation of the vortex in the both areas a third crucial step is added. There must be an appropriate method for coupling the sub-grid scale models with the results of direct numerical calculation process.

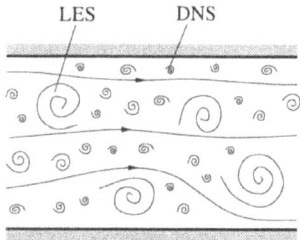

Fig. 4.4a: Schematic representation of a turbulent flow with large eddies (LES) and small eddies (DNS) according to [87].

Fig. 4.4b illustrates that the spectrum of turbulent kinetic energy. Depending on the algorithm used for direct numerical simulation of the larger vortices, they can be resolved by various numerically related effects. If the algorithm used allows no dissipation of energy, it will be used in the range of small length scales (Fig. 4.4a). The energy rising up can then impinge on larger vortex and leads to unrealistic results. Other algorithms lead to the conversion of kinetic energy into heat for small, yet resolved scales one (Fig. 4.4b). Extending this effect too much larger scales it results with excessive energy dissipation (Fig. 4.4b).

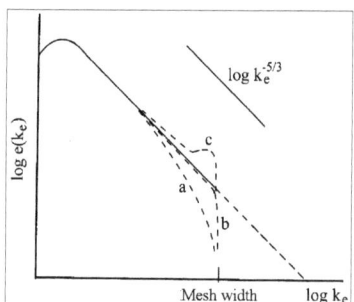

Fig. 4.4b: Influence of different algorithms on the calculated turbulent energy spectrum [80]: (a) the algorithm is to diffusive (b) MILES algorithm, (c) the algorithm blocks the energy dissipation.

To solve a problem in the transition between resolved and modeled scales to a Large Eddy Simulations filtering features are introduced. They suppress the unwanted numerical effect on the cut off of the grid; on the other hand, they serve as a basis for describing the transport processes beneath the grid size. Conventional LES does not solve the Navier-Stokes equations directly, but a secondary through a set of equations for filtered sizes. So, the balance size Φ consists of a large part $\overline{\Phi}$ numerically resolved and a small, non-resolved part Φ' [80]:

$$\Phi = \overline{\Phi} + \Phi'. \qquad (4.26)$$

The resolution is part of volume averaging, taking into account the filter function G [80]:

$$\overline{\Phi}(x,y,z) = \int_V G(x-x', y-y', z-z')\Phi(x',y',z',t)dx'dy'dz'. \qquad (4.27)$$

As a filter function, for example, top-hat filter or Gaussian filter can be used. The cell width below which the eddy is no longer resolved a user can chose freely. In general, for Large Eddy Simulation, however, a much finer time and space discretisation is required than for RANS modeling. As a maximum time step size can be a value of $t = 10^{-4}$ s [83]. The calculation of the temporal evolution of a particular size in a certain time interval of at least $t_{tot} = 20$ s is, however, with the current state of computer technology extremely time consuming. Large Eddy Simulations need boundary layers in an extremely fine grid. Therefore, the combinations of LES and unsteady RANS

(URANS) are designed to work on this problem. In the so called Scale Adaptive Simulation [83] is depending on the size of cells where LES switches to URANS. Ideally, the transient, anisotropic turbulence in the border areas through layers detached LES is directly calculated. The boundary layers are, however, with URANS models simulated. This allows the computing time to be significantly reduced. However, the results are dependent directly from the grid, the transfer point of the vortex from the boundary layer can with an inappropriate cell sizes be negatively affected.

Turbulent flows are characterized by eddies with a wide range of length and time scales.

The largest eddies are typically comparable in size to the characteristic length of the mean flow. The smallest scales are responsible for the dissipation of turbulence kinetic energy. It is possible, in theory, to directly resolve the whole spectrum of turbulent scales using an approach known as direct numerical simulation (DNS). No modeling is required in DNS. However, DNS is not feasible for practical engineering problems involving high Reynolds number flows. The cost required for DNS to resolve the entire range of scales is proportional to Re_t^3, where Re_t is the turbulent Reynolds number. Clearly, for high Reynolds numbers, the cost becomes prohibitive. In LES, large eddies are resolved directly, while small eddies are modeled. Large eddy simulation (LES) thus falls between DNS and RANS in terms of the fraction of the resolved scales. The rationale of LES can be summarized as follows:

- Momentum, mass, energy, and other passive scalars are transported mostly by large eddies.
- Large eddies are more problem-dependent. They are dictated by the geometries and boundary conditions of the flow involved.
- Small eddies are less dependent on the geometry, tend to be more isotropic, and are consequently more universal.
- The chance of finding a universal turbulence model is much higher for small eddies.

Resolving only the large eddies allows one to use much coarser mesh and larger times step sizes in LES than in DNS. However, LES still requires substantially finer meshes than those typically used for RANS calculations. In addition, LES has to be run for a sufficiently long flow-time to obtain stable statistics of the flow being modeled. As a result, the computational cost involved with LES is normally orders of magnitudes higher than that for steady RANS calculations in terms of memory (RAM) and CPU time. Therefore, high-performance computing (e.g., parallel computing) is a necessity for LES, especially for industrial applications. The following text gives details of the governing equations for LES, the sub-grid scale turbulence models, and the boundary conditions.

The governing equations employed for LES are obtained by filtering the time-dependent Navier-Stokes equations in either Fourier (wave-number) space or configuration (physical) space. The filtering process effectively filters out the eddies which scales are smaller than the filter width or grid spacing used in the computations. The resulting equations thus govern the dynamics of large eddies.

A filtered variable (denoted by an over bar) is defined by

$$\bar{j}(x) = \int_D j(x')G(x,x')dx' \tag{4.28}$$

where D is the fluid domain, and G is the filter function that determines the scale of the resolved eddies. The finite-volume discretization itself implicitly provides the filtering operation:

$$\bar{j}(x) = \frac{1}{V}\int_V j(x')dx', \quad x' \in v \tag{4.29}$$

where V is the volume of a computational cell. The filter function, $G(x,x')$, implied here is then

$$G(x,x')\begin{cases} 1/V, & x' \in v, \\ 0, & x' \text{ otherwise} \end{cases} \tag{4.30}$$

The LES capability is applicable to compressible flows. For the sake of concise notation, however, the theory is presented here for incompressible flows.

Filtering the Navier-Stokes equations, one obtains

$$\frac{\partial \rho}{\partial t} + \frac{\partial \rho}{\partial x_i}(\rho \bar{u}_i) = 0 \tag{4.31}$$

and

$$\frac{\partial}{\partial t}(\rho \bar{u}_i) + \frac{\partial}{\partial x_i}(\rho \bar{u}_i \bar{u}_j) = \frac{\partial}{\partial x_j}(\mu \frac{\partial \sigma_{ij}}{\partial x_j}) - \frac{\partial \bar{p}}{\partial x_i} - \frac{\partial \tau_{ij}}{\partial x_j} \tag{4.32}$$

where σ_{ij} is the stress tensor due to molecular viscosity defined by

$$\sigma_{ij} \equiv \left[\mu\left(\frac{\partial \overline{u}_i}{\partial x_j}+\frac{\partial \overline{u}_j}{\partial x_i}\right)\right]-\frac{2}{3}\mu\frac{\partial \overline{u}_i}{\partial x_i}\delta_{ij} \qquad (4.33)$$

and τ_{ij} is the sub-grid scale stress defined by

$$\tau_{ij} \equiv \overline{\rho u_i u_j} - \overline{\rho u_i}\,\overline{u_j}. \qquad (4.34)$$

The sub-grid scale stresses resulting from the filtering operation are unknown, and require modeling. The sub-grid scale turbulence models in CFX and FLUENT employ the Boussinesq hypothesis [83] as in the RANS models, computing sub-grid scale turbulent stresses from

$$\tau_{ij} - \frac{1}{3}\tau_{kk}\delta_{ij} = -2\mu_t \overline{S}_{ij} \qquad (4.35)$$

where μ_t is the sub-grid scale turbulent viscosity. The isotropic part of the sub-grid scale stresses τ_{kk} is not modeled, but added to the filtered static pressure term. $\overline{S}_{ij}$ is the rate of strain tensor for the resolved scale defined by

$$\overline{S}_{ij} \equiv \frac{1}{2}\left(\frac{\partial \overline{u}_i}{\partial x_j}+\frac{\partial \overline{u}_j}{\partial x_i}\right). \qquad (4.36)$$

For compressible flows, it is convenient to introduce the density-weighted (or Favre) filtering operator:

$$\Phi = \frac{\overline{\rho\Phi}}{\overline{\rho}}. \qquad (4.37)$$

A further detail in defining the coefficients can be found in [84].

The dynamic Smagorinsky-Lilly model is used in this work. Here, a sub-grid scale turbulent flux of a scalar, Φ, is modeled using s sub-grid scale turbulent Prandtl number by

$$q_j = -\frac{\mu_t}{\sigma_t}\frac{\partial \Phi}{\partial x_j} \qquad (4.38)$$

where q_j is the sub-grid scale flux.

Germano et al. [84] and subsequently Lilly [84] conceived a procedure in which the Smagorinsky model constant, Cs, is dynamically computed based on the information provided by the resolved scales of motion. The dynamic procedure thus obviates the need for users to specify the model constant Cs in advance. The details of the model

74   4 Some important topics of CFD used in this work

implementation in FLUENT and its validation can be found in [84]. The Cs obtained using the dynamic Smagorinsky-Lilly model varies in time and space over a fairly wide range. To avoid numerical instability, Cs is clipped at zero and 0.23 by default.

4.2.1.3 Scale Adaptive Simulations

SAS [83,89] is a hybrid model contains the Unsteady Reynolds Averaged Navier Stokes Equation (URANS) and the Large Eddy Simulation (LES), where URANS acts in the near of wall boundaries and LES in the remaining part of the domain. SAS is an improved URANS formulation, which allows the resolution of the turbulent spectrum in unstable flow conditions. SAS model dynamically adjusts to resolved structures in a URANS simulation, which results in a LES-like behavior in unsteady regions of the flow field. At the same time, the model provides standard RANS capabilities in stable flow regions.

Scale-Adaptive Simulation (SAS) allows the simulation of unsteady flows with both RANS and LES content in a single model environment. As SAS formulations use the von Karman length scale as a second external scale, they can automatically adjust to resolved features in the flow. As a result, SAS develops LES like solutions in unsteady regions, without a resort to the local grid spacing.

The concept of Scale-Adaptive Simulation (SAS) [83,89] allows the simulation of unsteady turbulent flows without the limitations of most Unsteady RANS (URANS) models. Contrary to standard URANS, SAS provides two independent scales to the source terms of the underlying two-equation model. In addition to the standard input in form of the velocity gradient tensor, $\partial u_i/\partial x_j$, SAS models compute a second scale from the second derivative of the velocity field. The resulting length scale is the well known von Karman length scale L_{vK}. The introduction of L_{vK} allows the model to react more dynamically to resolved scales in the flow field which cannot be handled by standard URANS models. As a result, SAS offers a single framework, which covers steady state RANS as well as LES regions, without an explicit switch in the model formulation. SAS therefore offers an attractive framework for many "multi-scale" flow problems encountered in industrial CFD. It provides a steady state (or mildly unsteady) solution in stable flow regions (like boundary layers), and unsteady structures in unsteady regions within a single model framework. SAS allows a breakdown of the large unsteady structures by adapting the turbulence model to the locally grid spacing [83,89].

The SAS approach represents a new class of the URANS models. Different from the conventional RANS formulations, the SAS model adjusts the turbulence length scale to the local flow inhomogeneities. As a measure of the local flow length scale, a classic boundary layer length scale introduced by von Karman $\kappa U'(y)/U''(y)$ is generalized for arbitrary three-dimensional flows. The von Karman length scale explicitly enters the transport equations of the turbulence model. The resulting model remains a RANS model, as it delivers proper RANS solutions for stationary flows and maintains these solutions through grid refinement. On the other hand, for flows with transient instabilities like those in the massive separation zones, the model reduces its Eddy viscosity according to the locally resolved vortex size represented by the von Karman length scale. The SAS model can under those conditions resolve the turbulent spectrum down to the grid limit and avoids RANS-typical single-mode vortex structure.

1. SAS modeling is based on the use of a second mechanical scale in the source/sink terms of the underlying turbulence model. In addition to the standard input from the momentum equations in the form of first velocity gradients (strain rate tensor, vorticity tensor) SAS models rely on a second scale, in the form of higher velocity gradients (typically second derivatives).

2. SAS models satisfy the following requirements:

   a. Provides proper RANS performance in stable flow regions.

   b. Allows the break-up of large unsteady structures into a turbulent spectrum.

   c. Provides proper damping of resolved turbulence at the high wave number end of the spectrum (resolution limit of the grid).

3. Functions (2a) and (2b) are achieved without an explicit grid or time step dependency in the model. Naturally, function (2c) has to be based on information on the grid spacing, other information concerning the resolution limit (dynamic LES model, etc.), or the numerical method (MILES damping etc.)

A complete description of the SST-SAS model will be published elsewhere due to space limitations. A compressed model formulation is provided here to leave more space for the test case results. The governing equations of the SST-SAS model differ from those of the SST RANS model by the additional SAS source term Q_{SAS} in the transport equation for the turbulence eddy frequency ω (Eq. 4.39a,b):

$$\frac{\partial(\rho k)}{\partial t} + \nabla \cdot (\rho U k) = P_k - \rho c_\mu k \omega + \nabla \cdot \left[\left(\mu + \frac{\mu_t}{\sigma_k}\right)\nabla k\right] \tag{4.39a}$$

$$\frac{\partial(\rho \omega)}{\partial t} + \nabla \cdot (\rho U \omega) = \alpha \frac{\omega}{k} P_k - \rho \beta \omega^2 + Q_{SAS} + \nabla \cdot \left[\left(\mu + \frac{\mu_t}{\sigma_\omega}\right)\nabla \omega\right] + (1-F_1)\frac{2\rho}{\sigma_{\omega 2}}\frac{1}{\omega}\nabla k \nabla \omega \tag{4.39b}$$

where $\sigma_{\omega 2}$ is the σ_ω value for the k-ε regime of the SST model.

$$Q_{SAS} = \max\left[\rho \xi_2 \kappa S^2 \left(\frac{L}{L_{VK}}\right)^2 - C \cdot \frac{2\rho k}{\sigma_\phi} \max\left(\frac{|\nabla \omega|^2}{\omega^2}, \frac{|\nabla k|^2}{k^2}\right), 0\right]. \tag{4.40}$$

More details about coefficients used in equations included in SAS model can be found in [83,89].

Adaptive scale simulation (SAS) model is advanced URANS (unsteady RANS) model which allows a realistic calculation of unsteady flows. Unlike standard URANS methods, which typically resolve only large scale structures, SAS allows a greater turbulence decay elements in a turbulent spectrum. SAS reaches the transient improvement in the field by taking into account the Karman length scale in the scaling of the equation turbulence model. This allows the model already resolved structures and avoids the smearing which is typical in the standard URANS methods. A detailed mathematical description can be found in CFX guideline [83].

4.2.2 Combustion models
4.2.2.1 Source terms of species on the basis of chemical reactions

In reactive flows for i–1 species for each species will one balance equation Eq. (4.12) be solved (Section 4.1.2). The changes of the species mass fractions Y_i due to chemical reactions are considered in a source term S_i. It follows for all first reactions with all species with the stoichiometric coefficients $v'_{i,r}$ and $v''_{i,r}$ of the species i on the reaction or product side in the reaction r:

$$\sum_{i=A,B,C,...}^{N_S} v'_{i,r} i = \sum_{i=A,B,C,...}^{N_S} v''_{i,r}. \qquad (4.41)$$

The source S_i, used to increase or decrease the mass fraction of a contributed species, is calculated as a sum of the changes due to all reactions:

$$S_i = M_i \sum_{i=A,B,C,...}^{N_S} (v'_{i,r} - v''_{i,r}) R_r. \qquad (4.42)$$

The reaction speed R_r can be calculated by using different models, for example, with an Eddy Break Up model [68,83], the Eddy Dissipation model [68,83] or a Finite Rate Chemistry model [83]. As a main part of this work is the modeling of combustion of JP-4 (jet propulsion) fuel. The chemical reactions for combustion of JP-4 fuel are calculated with three different approaches: multistep chemical reaction and two different flamelet models: the Laminar flamelet model with pdf, with 112 species and 800 chemical reactions and Pdf transport model for species with 20 species and 40 chemical reactions (Section 4.4.3.2).

4.2.2.2 The Eddy Dissipation Model

The Eddy dissipation model is based on the concept that chemical reaction is fast relative to the transport processes in the flow. When reactants mix at the molecular level, they instantaneously form products. The model assumes that the reaction rate may be related directly to the time required to mix reactants at the molecular level. In turbulent flows, this mixing time is dominated by the eddy properties and, therefore, the rate is proportional to a mixing time defined by the turbulent kinetic energy, k, and dissipation, ε.

$$\text{Rate} \propto \frac{\varepsilon}{k}. \qquad (4.43)$$

This concept of reaction control is applicable in many industrial combustion problems where reaction rates are fast compared to reactant mixing rates. In the Eddy Dissipation model, the rate of progress of elementary reaction k, is determined by the smallest of the two following expressions:

$$\text{Reactants limiter:} \quad R_k = A \frac{\varepsilon}{k} \min\left(\frac{c(I)}{v'_{kI}}\right) \qquad (4.44)$$

where c(I) is the molar concentration of component I and I only includes the reactant components.

Product limiter: $R_k = AB \dfrac{\varepsilon}{k} \left(\dfrac{\sum_P c(I)W_I}{\sum_P v''_{kI} W_I} \right)$ (4.45)

where P loops over all product components in the elementary reaction k. Details can be found in [83].

4.2.2.3 Flamelet model

Non-premixed flame is characterized in general by a diffusive transportation of fuel and oxidizer through the flame zone, which is slow compared to the chemical reaction [68,83,90]. If the molecular mixture occurs below the level of the smallest vortices, the combustion takes place then at the surface of theoretically infinite thin layers, and layers are smaller than the Kolmogorov length. These one dimensional, laminar flame fronts will hence refer to flamelets. The influence of turbulence, expressed in the folding and stretching of the flamelets from the chemical structure is however not affected. Because of that the mixture process strongly depends on the diffusion, it is easier to describe what happens by using a particular diffusion coefficient for each scalar. Then runs the mixture of all species equally fast, so only a single variable in the mixture is pursued. As chemical species are formed or consumed in reactions, it is easier to consider the elements which mass fractions remain unchanged. So, in a two stream system of the mixture fraction with mass fractions Z of element j it follows [68,83]:

$$\xi = \dfrac{Z_j - Z_{j,2}}{Z_{j,1} - Z_{j,2}}$$ (4.46)

with Z defined as

$$Z_j = \sum_{i=1}^{N_S} \dfrac{a_{i,j} M_j}{M_i} Y_i$$ (4.47)

and the molar masses M and mass fractions Y of the species i. The indices 1 and 2 refer to the material flow. Due to that by definition all the diffusion coefficients are equal; there are no changes in the two streams in the relationship between the elements. It is easy to show that the mass fraction regardless to the considered element is [68]:

$\xi_1 = \xi_2 = ... = \xi_{Ns}$. (4.48)

Boundary conditions applies $\xi = 1$ in the fuel stream and $\xi = 0$ in oxidant stream. The mixture fraction can as the mass fraction of the material, be interpreted as the stream 1. The mixture fraction of material in a stream 2 is than $1 - \xi$. After summation of all species equations the equation for the mass fraction is obtained [68,83]:

$$\frac{\partial(\rho\xi)}{\partial t} + \nabla(\rho\xi) = \nabla(\rho D \nabla \xi). \quad (4.49)$$

The mass fraction is independent of the chemical reactions, because it refers to chemical elements, which are not consumed or formed by chemical reactions. Since the balance equation does not contain a source term, ξ is also as scalar conserved scalar [83]. Accounting to that the ratio of material to heat transport can be predicted in connection to the Lewis number

$$Le = \frac{\lambda}{\rho c_p D} \quad (4.50)$$

which is equal to 1 in the system without heat loses. To describe the situation on one plane a stoichiometric composition $\xi = \xi_{st}$ will a new, locally defined coordinate system introduce with coordinates x_2 and x_3 within the reaction zone. The x_1 coordinate is perpendicular to the reaction zone and can be defined under the assumption that these infinite thin zone is presented through the mixture ξ [83,90]. Therefore, the following transformation leads to:

$$\frac{\partial}{\partial t} = \frac{\partial}{\partial \tau} + \frac{\partial \xi}{\partial t}\frac{\partial}{\partial \xi}, \quad (4.51)$$

$$\frac{\partial}{\partial x_1} = \frac{\partial \xi}{\partial x_1}\frac{\partial}{\partial \xi}, \quad (4.52)$$

$$\frac{\partial}{\partial x_k} = \frac{\partial}{\partial \xi_k} + \frac{\partial \xi}{\partial x_k}\frac{\partial}{\partial \xi}. \quad (4.53)$$

One example is calculation of temperature transformation. For the temperature equation e.g. in ANSYS CFX follows [83]:

$$\rho\frac{\partial T}{\partial t} + \rho u_\alpha \frac{\partial T}{\partial x_\alpha} - \frac{\partial}{\partial x_\alpha}\left(\rho D \frac{\partial T}{\partial x_\alpha}\right) = \sum_{k=1}^{r} \frac{Q_k}{c_p}\omega_k + \frac{q_R}{c_p} + \frac{1}{c_p}\frac{\partial p}{\partial t}. \quad (4.54)$$

The three terms on the right side represent chemical reactions, the radiation and the transient pressure. The latest effect with rapid pressure changes such as piston engines. Since the flamelets are very thin, only the gradient normal to the surface of

4 Some important topics of CFD used in this work

the stoichiometric mixture has great values. This is taken into account in the terms, the second derivative in relation to the mixture fraction ξ are neglected to simplify the equation [83]:

$$\frac{\partial T}{\partial t} + \frac{\chi_{st}}{2}\frac{\partial^2 T}{\partial \xi^2} = \frac{1}{\rho c_p}\sum_{i=1}^{N_S} h_i \dot{m}_i + \frac{q_R}{c_p} + \frac{1}{c_p}\frac{\partial p}{\partial t}. \tag{4.55}$$

One important parameter in Eq. (4.55) is scale dissipation rate which represents an influence of the flow field to the reaction zone and it is given by

$$\tilde{\chi} = C_\chi \frac{\tilde{\varepsilon}}{k} \xi''^2. \tag{4.56}$$

The reaction zone is determined by the flow field stretched by diffusion. At a critical scalar dissipation χ_{cr} is the reaction zone so thin that the heat loss through conduction can not longer be compensated. The temperatures decrease strongly, locally and the flame is extinguished at this point. This is of great importance, because this parameter is ensured that a turbulent flame can be described by an ensemble of one dimensional flame. The scalar dissipation χ_{st} and the pressure p are flamelet parameters [90]. That the influences of turbulence in Flamelet model can be taken into account a two balance equations are needed. The first is required for the Favre averaged mixture fraction [83,84]:

$$\frac{\partial(\bar{\rho}\tilde{\xi})}{\partial t} + \nabla(\bar{\rho}\bar{u}\tilde{\xi}) = \nabla\left[\bar{\mu} + \frac{\mu_t}{\sigma_\xi}\nabla\tilde{\xi}\right]. \tag{4.57}$$

The two additional terms represent the production and dissipation of the variance term. The dissipation is included in CFX by the empirical correlation:

$$\tilde{\chi} = C_\chi \frac{\tilde{\varepsilon}}{k} \xi''^2. \tag{4.58}$$

with the constants $\xi = \xi''^2 = 0.9$, $C_\chi = 2.0$. Favre averaged composition of the fluid is a function of Favre averaged mixture fraction, its variance and the scalar dissipation rate:

$$\tilde{Y}_i = \tilde{Y}_i(\tilde{\xi}, \tilde{\xi}''^2, \tilde{\chi}) = \int_0^1 \tilde{Y}_i(\tilde{\xi}, \tilde{\chi}_{st}) P_{\tilde{\xi}, \tilde{\xi}''^2}(\xi)\partial\xi. \tag{4.59}$$

The function P is a probability density function (pdf). It is used in the statistical description of the flames properties. The probability density function pdf of a fluid gives the probability that the fluid in a location r have a density between ρ and ρ+dρ, a speed in x direction between u_x and $u_x + du_x$, in y direction between u_y and $u_y + \Delta u_y$ and in z direction between u_z and $u_z + du_z$, a temperature in the range between T and T + dT and local mass fractions between Y_i and $Y_i + dY_i$ [68]:

$$P(\rho, u_x, u_y, u_z, Y_1, ..., Y_{N_S-1}, T; \vec{r}) d\rho du_x du_y du_z dY_1, ..., dY_{N_S-1}, dT . \tag{4.60}$$

The probability density function used in this work is based on empirical data. In this way a statistical independence of each variable is adopted regard to the pdf of one variable. In this way, the multidimensional pdf is formulated as one dimensional pdf [68]:

$$P(\rho, u_x, u_y, u_z, Y_1, ..., Y_{N_S-1}, T; \vec{r}) = P(\rho)P(u_x)P(u_y)P(u_z)P(Y_1), ..., P(Y_{N_S-1})P(T) .$$

$$\tag{4.61}$$

As individual variables are not independent of each other, additional correlations between them are introduced. The other one dimensional pdf from experiments is used to determine which the crucial advantage in the above separation lies.

Fig. 4.5 shows an example a schematic representation of probability density functions for the mass fraction of the fuel mixture in a turbulent layer, such as on the edge of a pool of flame can occur. On the edge of the mixing layer is the probability to find a pure fuel or a pure air relatively high. Within the mixing process, the probability to found, one of the two components unmixed, is getting lower. Nevertheless, for a fixed locations pdf one of the two components unmixed to be found is equal to zero, because of fluctuations the local limits of the reaction zones are always shifting. For analytical description of such one dimensional pdf different application functions such as Gaussian function or β-function can be used. By inserting the mixture fraction can with the help of the β-function Eq. (4.59) be integrated. The implementation of this calculation step in each cell, for the each iteration will make a substantial computational effort. Therefore, the concept of look-up table was introduced. With this approach a preprocessor will make integration depending on the mixture fracture. For the various mixture fractions resulting densities, velocities, temperatures and concentrations of species are contained in the so-called look-up table, so that during the actual simulation the compute-intensive integration is avoided. Instead, the corresponding values for the flow field and calculated mixture fraction are readable directly from the table. Through these savings will consideration of numerous species and their elementary reactions with included empirical certain reaction speeds be possible.

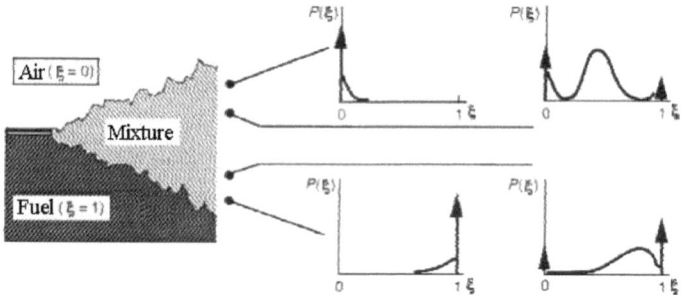

Fig. 4.5: Schematic representation of probability density functions for the mass fraction of the fuel mixture in a turbulent layer [68].

In the case of the calculation made here by using JP-4 (Jet propulsion fuel) combustion, a reaction mechanism according [83] is used, in which 112 species in 800 elementary reactions react one with another. A simplified flamelet model with 20 species and 40 elementary reactions can be found in [84].

4.2.3 Radiation models

4.2.3.1. Photometric sizes and radiation balance equation

While the heat conduction in solids is the only type of heat transport, in turbulent flows plays a minor role. Here, dominate depending on the temperature either convection or thermal radiation. The different meanings of the thermal radiation can be recognized by the Stefan-Boltzmann Law, according which the surface emissive power is dependent on the fourth power of temperature [85]:

$$SEP(T) = \sigma T^4 . \quad (4.62)$$

Thus, contribution of radiation at relatively low temperatures are only slightly in exchange of energy, at high temperatures, however, it significantly determines the heat transfer. Therefore, the consideration of radiation in here simulated pool flames and combustion processes generally have a crucial importance. Any matter, whether gaseous, liquid or solid, emit at any temperature $T > 0$ thermal radiation in the wavelength range of 0.1 ηm ≤ λ ≤ 100 ηm. The visible wavelength range makes only a small part of the total energy, the largest part is emitted in the infrared range (Fig. 4.6).

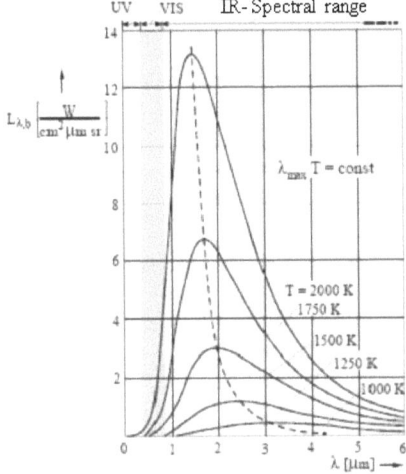

Fig. 4.6: Spectral radiance distribution of a black radiator [85].

Incidence of photons in fluids or solids can produce different effects. By absorption of a photon its energy is transferred to the fluid or solid. Gases show unlike solids due to their different transitions discrete energy absorption and emission spectra. The atoms of a gas are electronically driven, quantized states, in molecules exist discrete vibration and rotation states. Accordingly, absorption and emission of a gas are strongly wavelength dependent. In contrast, usually it is assumed that the radiant energy of impermeable body depends only on temperature and physical characteristics of the body. Their emission spectrum is independent on the wavelength of the incident radiation. This assumption is based on the fact that usually the entire radiation energy absorbed is converted to the internal energy with an equilibrium distribution. Besides that absorption and emission can still take a place. Famous examples include the scattering of particles such as dust clouds or liquid droplets in mist. Collisions are

84   4 Some important topics of CFD used in this work

scattering of photons with other particles, where the photons are diverted. During clashes elastic energy remains unchanged, at inelastic collisions; however, the energy is usually reduced. If the scattering is direction preferred, it is an anisotropic scattering. In an isotropic scattering a uniform distribution of the scattered photons occurs. To describe the radiation the radiometry various sizes are defined. The accounting of the radiation is usually based on the radiance L. It is defined as the radiant flux Φ_{rad} that in an differential wavelength range $d\lambda$ projected in direction from the vertical to ϑ, φ, emits to a radiant surface element $dA_P = dA_1 \cos \vartheta$ in the space angle [85]:

$$L(r,s,\lambda) = \frac{d^3 \Phi_{rad}}{dA_1 \cos\vartheta \, d\Omega \, d\lambda} = \frac{dI(\lambda)}{dA_1 \cos\vartheta} \qquad (4.63)$$

I is called here radiant intensity. The geometric relationships are illustrated in Fig. 4.7. The radiation flux Φ_{rad} is defined as

$$\Phi_{rad} = \frac{dQ}{dt} \qquad (4.64)$$

with that by photon transported radiation energy Q. The product of the absolute temperature of a black body T and wavelength λ_{max} of maximum surface emissive power is constant and gives the Wien's displacement law [85]:

$$\lambda_{max} T = C_2/4.97 = 2.898 \cdot 10^{-3} \, mK. \qquad (4.65)$$

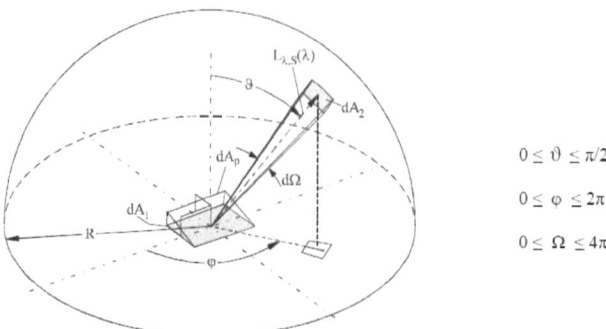

Fig. 4.7: Definition of radiance by geometric means [85].

The balance equation for the radiance is taking into account the possible effects [83,85]:

$$dL_\lambda(\vec{r},s) = \underbrace{-(k_{abs,\lambda} + k_{scat,\lambda})L_\lambda(\vec{r},s)}_{\substack{\text{weakening due to}\\\text{absorption and}\\\text{scattering}}} + \underbrace{k_{abs,\lambda} L_{S,\lambda}(\vec{r},\vec{s},T)}_{\substack{\text{weighted}\\\text{blackbody}\\\text{radiation}}} + \underbrace{\frac{k_{scat,\lambda}}{4\pi}\int dL_\lambda(r,s)\psi(\vec{s},\vec{s}',\Omega)d\Omega}_{\text{entered radiation}} + \underbrace{S_{L_\lambda}}_{\substack{\text{heat}\\\text{sources}}}$$

(4.66)

Accordingly, the amendment of the spectral radiance along the beam path l at the place r in the direction $\vec{s}$ is equal to weakening due to absorption and scattering with the coefficient k_{abs} and k_{scat}, by increasing the intensity in the direction $\vec{s}$ and issued with the coefficients k_{abs} weighted blackbody radiation L_B ($r,\vec{s}$,T), and the rest from all directions $\vec{s}'$ entered radiation. The function $\Psi(\vec{s},\vec{s}',\Omega)$ stands for the probability that the beam $\vec{s}'$ at the beam thickness $d\Omega$ causes an increase in the direction $\vec{s}$. Furthermore, additional heat sources S_{L_λ} are taken into account. The modeled pool flames are characterized by high temperatures and a very high soot particle concentration. The vast majority of the radiation is emitted from hot soot particles. Therefore, as a good approximation can be assumed that the gas/soot mixture is a gray medium. With the assumption that the radiation intensity over all wavelengths is equally distributed, the intensive compute consideration of the various runs of existing gas species is eliminated. In addition to the radiance are often given a further two sizes. The hemispherical surface emissive power SEP is the radiant flux in one differential wavelength range $d\lambda$ of a surface element dA_1, integrated across a space angle $2\pi\Omega_0$ of half of the sphere:

$$SEP(r,\vec{s},\lambda) = \frac{d^2\Phi_{rad}}{dA_1 d\lambda}.$$ (4.67)

The irradiance E does not refers in contrast to the radiance and surface emissive power to the radiative surface but to the receiving area element. The spectral irradiance is defined as:

$$E(r,\lambda) = \frac{d^2\Phi_{rad}}{dA_2 d\lambda}.$$ (4.68)

4.2.3.2 Discrete Ordinate

Beside the Monte Carlo simulations is also the Discrete Ordinate radiation model in modeling of pool flames enforced. It is included in e.g. ANSYS FLUENT software [84]. The Discrete Ordinate (DO) radiation model solves the radiative transfer equation (RTE) for a finite number of discrete solid angles, each associated with a vector direction $\vec{s}$ fixed in the global Cartesian system (x, y, z). The fineness of the angular discretization is controlled by user, analogous to choosing the number of rays in a case of Discrete Transfer Radiation Model (DTRM) included in e.g. ANSYS CFX [83]. Unlike the DTRM, however, the DO model does not perform ray tracing. Instead, the DO model transforms Eq. 4.66 into a transport equation for radiation intensity in the spatial coordinates (x, y, z). The DO model solves for as many transport equations as there are directions $\vec{s}$. The solution method is identical to that used for the fluid flow and energy equations. Two implementations of the DO model are available in FLUENT: uncoupled and (energy) coupled. The uncoupled implementation is sequential in nature and uses a conservative variant of the DO model called the finite-volume scheme [84], and its extension to unstructured meshes [83,84]. In the uncoupled case, the equations for the energy and radiation intensities are solved one by one, assuming prevailing values for other variables. Alternatively, in the coupled ordinates method (or COMET) [84], the discrete energy and intensity equations at each cell are solved simultaneously, assuming that spatial neighbors are known. The advantages of using the coupled approach are that it speeds up applications involving high optical thicknesses and/or high scattering coefficients. Such applications slow down convergence drastically when the sequential approach is used.

The DO model considers the radiative transfer equation (RTE) in the direction $\vec{s}$ as a field equation. Thus, Eq. (4.66) is written as:

$$\nabla \cdot (I(\vec{r},\vec{s})\vec{s}) + (a + \sigma_s)I(\vec{r},\vec{s}) = an^2 \frac{\sigma T^4}{\pi} + \frac{\sigma_s}{\pi} \int_0^{4\pi} I(\vec{r},\vec{s}')\Phi(\vec{s} \cdot \vec{s}')d\Omega' . \tag{4.69}$$

FLUENT also allows the modeling of non-gray radiation using a gray-band model. The RTE for the spectral intensity $I_\lambda(\vec{r},\vec{s})$ can be written as:

$$\nabla \cdot (I_\lambda(\vec{r},\vec{s})\vec{s}) + (a_\lambda + \sigma_s)I_\lambda(\vec{r},\vec{s}) = a_\lambda n^2 I_{b\lambda} + \frac{\sigma_s}{4\pi} \int_0^{4\pi} I_\lambda(\vec{r},\vec{s}')\Phi(\vec{s} \cdot \vec{s}')d\Omega' . \tag{4.70}$$

4.2 Sub-models in fire modeling 87

Here λ is the wavelength, a_λ is the spectral absorption coefficient, and $I_{b\lambda}$ is the blackbody intensity given by the Planck function. The scattering coefficient, the scattering phase function, and the refractive index n are assumed independent of wavelength.

The total intensity $I(\vec{r},\vec{s})$ in each direction $\vec{s}$ at position $\vec{r}$ is computed using

$$I(\vec{r},\vec{s}) = \sum_k I_{\lambda k}(\vec{r},\vec{s})\Delta\lambda_k . \tag{4.71}$$

4.2.3.3 Monte Carlo

The Monte Carlo approach is widely used in the following cases:

1. For a solid media in geometry model in which radiation plays a role, only the Monte Carlo model is suitable.

2. For media interfaces in the model must on both sides the same radiation interface model used. In the case of a solid-liquid separation and a solid-solid state separation is only the Monte Carlo model suitable.

3. During the simulation with several solids each radiation model can be chosen independently.

The Monte Carlo model assumes that the intensity is proportional to the differential angular flux of photons assuming the radiation field as a photon gas. For this gas, K_a is the probability per unit length that a photon is absorbed at a given frequency. Therefore, the mean radiation intensity, I is proportional to the distance travelled by a photon in unit volume at r, in unit time. Similarly q_v^R is proportional to the rate of incidence of photons on the surface at r, since volumetric absorption is proportional to the rate of absorption of photons. By following a typical selection of photons and tallying, in each volume element, the distance travelled, the mean total intensity is obtained. By following a typical selection of photons and tallying, in each volume element, the distance times the absorption coefficient, the mean total absorbed intensity is obtained. By also tallying the number of photons incident on a surface and this number times the emissivity, the mean total radiative flux and the mean absorbed flux is obtained [83,84]. In the general version of the Monte Carlo model the coordinates of the starting points and the direction of radiation throughout the computing field are arbitrarily defined. When included in the code ANSYS CFX and FLUENT, the Monte Carlo model gives a certain amount of radiation referring to a geometric center of the cells with fixed starting point and a fixed orientation. This simplification is possible if the cells are small enough to assume that the radiation on nearly all solid isotropic dΩ is emitted. Each beam goes from its launch point until it leaves the computing field or expires. The sum of the energy absorbed of each cell results with absorbed energy of all radiation:

$$k_i(T) = 0.1 \cdot 10^5 s^{-1} e^{-21100/T}. \qquad (4.72)$$

The source of the energy conservation equation of a cell resulting from the difference between absorbed and emitted energy:

$$\frac{\partial(\rho Y_s)}{\partial t} + \nabla(\rho u Y_s) = \nabla\left(\frac{\mu_t}{Sc_t}\nabla Y_s\right) + S_{Y_s}. \qquad (4.73)$$

The calculation of the beams reflectance on the wall in one direction corresponds to the traditional Monte Carlo approach for an ideal diffusive reflector.

For a Monte Carlo calculation the total surface current and absorbed photon flux is printed, these figures should sum to unity and can therefore be used as another measure of the accuracy of the calculation. Next is the number of histories computed. The computational effort is estimated to trace a photon (ray) to the next event (surface) [83,84].

4.2.4 Soot models

Soot formation models, Magnussen, Lindstedt and Tesner are empirically based, approximate models of the soot formation process in combustion systems. The detailed chemistry and physics of soot formation are quite complex and are only approximated in the models used by ANSYS CFX and FLUENT.

4.2.4.1 Magnussen soot model

In the Magnussen soot model (Magnussen and Hjertager [92], it is assumed that soot is formed from a gaseous fuel in two stages, where the first stage represents formation of radical nuclei, and the second stage represents soot particle formation from these nuclei. Transport equations are solved for the specific concentration of radical nuclei X_N (mol/kg), and for the soot mass fraction $\tilde{Y}_s$ (kg/kg):

$$\frac{\partial(\bar{\rho}\tilde{X}_N)}{\partial t} + \frac{\partial(\bar{\rho}\tilde{u}_j\tilde{X}_N)}{\partial x_j} = \left\{\left(\bar{\mu} + \frac{\mu_t}{Pr_t}\right)\frac{\partial\tilde{X}_N}{\partial x_j}\right\} + \tilde{S}_{nuclei,f} + \tilde{S}_{nuclei,c}, \qquad (4.74)$$

$$\frac{\partial(\overline{\rho}\tilde{Y}_s)}{\partial t} + \frac{\partial(\overline{\rho}\tilde{u}_j\tilde{Y}_s)}{\partial x_j} = \left\{\left(\overline{\mu} + \frac{\mu_t}{Pr_t}\right)\frac{\partial \tilde{Y}_s}{\partial x_j}\right\} + \tilde{S}_{soot,f} + \tilde{S}_{soot,c}. \qquad (4.75)$$

The modeling procedure can be grouped into three independent parts:

1. Formation of nuclei and soot particles following the models models of Tesner et al. [93]
2. Combustion of nuclei and soot particles
3. Magnussen's Eddy Dissipation Concept (EDC) for modeling the effect of turbulence on mean reaction rates.

The soot can be used in either single phase or multiphase flow (MPF) configurations. In multiphase calculations, however, the soot variables cannot be a separate phase but must be assigned to one of the fluids. Formation of nuclei and soot particles is computed following the empirical models of Tesner [93]. The source terms are formulated in terms of particle number concentrations for nuclei:

$$C_N = \rho \, A \, X_N, \qquad (4.76)$$

and soot particles:

$$C_s = \rho \frac{Y_s}{m_P}, \qquad (4.77)$$

where $A = 6.02214199 \cdot 10^{23}$ is Avogardo number and $m_P = \rho_{soot}\pi d^3/6$ is the mass of a soot particle ρ_{soot} and d are the density and the mean diameter of the soot particles, respectively. Details can be found in [83,84].

4.2.4.2 Lindstedt soot model

Lindstedt model leads to soot modeling in two sizes, for one so-called particle density N, the number of particles per indicate mass of the mixture, and continue to soot mass fraction Y_s [94,95]. For both sizes an additional balance equation will be introduced:

$$\frac{\partial(\rho N)}{\partial t} + \nabla(\rho u N) = \nabla(\frac{\mu_t}{Sc_t}\nabla N) + S_N \qquad (4.78)$$

$$\frac{\partial(\rho Y_s)}{\partial t} + \nabla(\rho u Y_s) = \nabla(\frac{\mu_t}{Sc_t}\nabla Y_s) + S_{Y_s}. \qquad (4.79)$$

The determination of the source terms S_N and S_{Y_S} is described below. Experiments show that acetylene has an important role in the various soot formation processes (Section 2.5.3). For the first acetylene molecule PAHs can be built later, they are partially involved in HACA mechanism. Accordingly, Lindstedt describes the nucleation and growth of particles surfaces depending on the acetylene concentration. In response to the term, he formulated the following equation: $C_2H_2 \rightarrow 2C(s) + H_2$. In the reaction rate constant must be that one which received high activation temperature, but it should be noted that the surfaces of older soot are less reactive than the freshly formed particles. The possibility of introducing an additional factor to describe the temporary declination Lindstedt signs as too complex and uncertain, so he decides to make the following assumptions. Experiments show that the initial growth of interface is directly connected with the concentration of acetylene and that combination can be brought [96]. Furthermore, less than 10% of the total soot mass will be formed by the starting soot particles [97]. As the starting particles Lindstedt has chosen, therefore, very small particles of 100 carbon atoms and a particle diameter of $d_s = 1.24$ nm. Since for the calculation of the soot area the individual reaction rates are used with the critical particle size. Variations of particle size, according to Lindstedt, are in a range $1 \text{ nm} \leq d_s \leq 10 \text{ nm}$. The activation temperature for the above reaction is chosen to be 21100 K lower than that chosen by other authors, only to get suitable description of the launch of the initial particle surface growth within its model. The response rate for the formation of the start particles is then given as

$$R_1 = k_1(T) c(C_2H_2). \tag{4.80}$$

with $S_{Y_s} = M_s (2R_1 + 2R_2 - R_3)$. \hfill (4.81)

The preexponential factor Lindstedt determine from calculation of acetylene flame and compare with the measurements.

The second response to the increase in soot mass fractions leads to the deposition of acetylene on the surface of the soot particles formulated as:

$C_2H_2 + nC(s) \rightarrow (n+2)C(s) + H_2$.

The speed of reaction is directly proportional to acetylene concentration and to soot particles surface:

$$R_2 = k_2(T) f(S) c(C_2H_2). \tag{4.82}$$

The surface of all soot particles S will be calculated in m^2/m^3:

$$S = \pi(d_s^2)(\rho N). \tag{4.83}$$

By using particle diameter [92] in Eq. (4.83) for the surface is given:

$$S = \pi \left(\frac{6}{\pi \rho_s} \frac{1}{N} Y_s \right)^{1/3} (\rho N). \tag{4.84}$$

For the density of soot particles a value $S = 2000$ kg/m^3 is used. A linear dependence of acetylene composition in the soot particles from the surface of which would be the declining response rate of older particles do not reflect any more and therefore leads to too high soot mass fraction. Therefore, Lindstedt makes a simplified assumption that the active sides of the particles proportional to the square root of the entire soot surface. This assumption is convicted in Eq. (4.83). The molar mass of $M_s = 12,011$ g/mol and the concentration of particles is calculated according to

$$c(C(s)) = \rho Y_s / M_s. \tag{4.85}$$

In the constant flow speed in Eq. (4.84) with a certain activation temperature according to Vandsburger [98] and preexponential factor according to Lindstedt it follows:

$$k_2(T) = 0.6 \cdot 10^4 \, m^{1/2} s^{-1} e^{-12100/T}. \tag{4.86}$$

The oxidation of soot particles can in principle be made with various approaches. For example, while Fenimore and Jones [70] start from an oxidation by OH radicals, Liu [99] consider in addition oxidation by oxygen. However the Lindstedt describes the soot oxidation, as well as Lee et al. [100] by the reaction with oxygen: $C(s) + 0.5O_2 \rightarrow CO$. Besides the oxygen concentration, also for this reaction step is soot surface important.

$$R_3 = k_3(T) S c(O_2). \tag{4.87}$$

With the influence on the particle density the agglomeration or coagulation of soot particles is formulated as:

$$nC(s) \rightarrow C_n(s). \tag{4.88}$$

For the speed of reaction from the Lindstedt model follows:

$$R_4 = 2 C_a d_s^{1/2} \left(\frac{6 \kappa T}{\rho_s} \right) (\rho N), \tag{4.89}$$

where κ is Boltzmann constant, C_a agglomeration rate. Here the value of $C_a = 9$ is used.

From the above reaction the source for the balance equation of the particle density N is:

$$S_N = \frac{2}{C_{min}} N_A R_1 - k_4(T)c(C(s))^{1/6}(\rho N)^{11/6}, \qquad (4.90)$$

Where the number of carbon atoms is defined $C_{min} = 100$ (s.o.), N_A is the Avogadro constant. The source for the balance equation for soot mass fraction is calculated according to the following equation:

$$\Phi_P = \Phi_{PU} + \beta \nabla \phi \Delta \vec{v}. \qquad (4.91)$$

From the described reaction mechanism follows that by soot oxidation although soot mass fraction declines, the number of particles or particle density, however, remains unchanged. Simplification, however, provides a well-to-use mechanism for the purpose of this work as accurate enough to be considered.

4.2.4.3 Tesner soot model

The two-step Tesner model [84,93] predicts the generation of radical nuclei and then computes the formation of soot on these nuclei. The combustion of the soot (and particle nuclei) is assumed to be governed by the Magnussen combustion rate [83,84,92].

The transport equations are solved for two scalar quantities: the soot mass fraction (Eq. 4.92) and the normalized radical nuclei concentration:

$$\frac{\partial(\bar{\rho}\tilde{X}_N^*)}{\partial t} + \nabla(\bar{\rho}\tilde{u}_j \tilde{X}_N^*) = \nabla\left(\frac{\mu_t}{Pr_{t,N}} \nabla \tilde{X}_N^*\right) + R_N^* \qquad (4.92)$$

where

$\tilde{X}_N^*$ = normalized radical nuclei concentration (particles $\cdot$ 10^{-15}/kg)

$Pr_{t,N}$ = turbulent Prandtl number for nuclei transport

R_N^* = normalized net rate of nuclei generation (particles $\cdot$ 10^{-15}/m^3 – s).

In these transport equations, the rates of nuclei and soot generation are the net rates, involving a balance between formation and combustion.

The two-step model computes the net rate of soot generation, R_{soot}, in the same way as the single-step model, as a balance of soot formation and soot combustion:

$$R_{soot} = R_{soot,form} - R_{soot,comb}. \qquad (4.93)$$

In the two-step model, however, the rate of soot formation, $R_{soot,form}$ depends on the concentration of radical nuclei, c_{nuc}:

$$R_{soot,form} = m_s(\alpha - \beta N_{soot})c_{nuc} \tag{4.94}$$

where

m_s = mean mass of soot particle (kg/particle)

N_{soot} = concentration of soot particles (particles/m^3)

c_{nuc} = radical nuclei concentration = ρb_{nuc} (particles/m^3)

a = empirical constant (s^{-1})

b = empirical constant (m^3/particle-s)

The rate of soot combustion, $R_{soot,comb}$ is computed in the same way as for the single-step model, using (Eq. 4.93).

The default constants for the two-step model are for combustion of acetylene (C_2H_2). These values should be modified for other fuels, since the sooting characteristics of acetylene are known to be different from those of saturated hydrocarbon fuels.

The net rate of nuclei generation in the two-step model is given by the balance of the nuclei formation rate and the nuclei combustion rate:

$$R^*_{nuc} = R^*_{nuc,form} - R^*_{nuc,comb} \tag{4.95}$$

where

$R^*_{nuc,form}$ = rate of nuclei formation (particles · 10^{-15}/m^3-s)

$R^*_{nuc,comb}$ = rate of nuclei combustion (particles · 10^{-15}/m^3-s)

The rate of nuclei formation, $R^*_{nuc,form}$ depends on a spontaneous formation and branching process, described by

$$R^*_{nuc,form} = \mu_0 + (f - g)c^*_{nuc} - g_0 c^*_{nuc} N_{soot} \tag{4.96}$$

$$\mu_0 = a^*_0 c^*_{fuel} e^{-E/RT} \tag{4.97}$$

where

c^*_{nuc} = normalized nuclei concentration (ρb^*_{nuc})

$a^*_0 = a_0/10^{15}$

a_0 = pre-exponential rate constant (particles/kg–s)

c_{fuel} = fuel concentration (kg/m3)

f − g = linear branching − termination coeffcient (s−1)

g_0 = linear termination on soot particles (m³/particle–s)

Note that the branching term, $(f - g)c^*_{nuc}$, in Eq. 4.96 is included only when the kinetic rate, μ_0, is greater than the limiting formation rate (10^5 particles/m³–s, by default).

The rate of nuclei combustion is assumed to be proportional to the rate of soot combustion:

$$R^*_{nuc,comb} = R_{soot,comb} \frac{b^*_{nuc}}{Y_{soot}} \qquad (4.98)$$

where the soot combustion rate, $R_{soot,comb}$ is given by Eq. 4.93.

Additional inputs include the stoichiometry of the fuel and soot combustion and (for the two-step model only) the average size (diameter) and density of the soot particles used to compute the soot particle mass, m_s, in Eq. 4.94 for the two step model. Stoichiometry for soot combustion is the mass stoichiometry, ν_{soot}, which computes the soot combustion rate in both soot models. The model assumes that the soot is pure carbon and that the oxidizer is O_2. The default value supplied by ANSYS FLUENT [84] is for combustion of propane (C_3H_8) by oxygen (O_2).

4.3 ANSYS CFX and FLUENT software
4.3.1 Discretisation methods and solution algorithms
4.3.1.1 Finite volume method

The concept of discretization assumes, in the numeric, the transfer of a continuous function in one function, which only eventually many points are considered. As discussed in the previous chapters, balance equations can not analytically be solved, the differential quotient by a difference in discretization quotient must be transferred. As part of this work a very widespread method of finite volume is applied, i.e. the balance area is to maximize the number of divided control volumes. This creates a computational grid with discrete nodes. A part of a two-dimensional mesh or grid is shown in Fig. 4.8. This is a reassembling grid, which means that not all sizes have the same grid points. The vector sizes such as the flow speeds are sent to the cell edge view, the scalar variables such as pressure are saved in the cell centers. Grids are often shifted from numerical stability reasons. The differential equations are numerically considered only to the discrete points. According to the definition of derivation of a differential function

$$\frac{d\Phi}{dx} = \lim_{\Delta x \to 0} \frac{\Phi_{x+\Delta x} - \Phi_x}{\Delta x}, \tag{4.99}$$

the differential operator at the grid point P is approximated by a differential operator [101]:

$$\left(\frac{d\Phi}{dx}\right)_P = \frac{\Phi_E - \Phi_P}{\Delta x}. \tag{4.100}$$

In this example a so called right side difference is used, also the left side difference is possible here [88]

$$\left(\frac{d\Phi}{dx}\right)_P = \frac{\Phi_P - \Phi_W}{\Delta x}, \tag{4.101}$$

or also a central difference [101]

$$\left(\frac{d\Phi}{dx}\right)_P = \frac{\Phi_E - \Phi_W}{2\Delta x}. \tag{4.102}$$

The central difference is assigned by square order error $(1/(\Delta x)^2)$, i.e. by half of the increment will discretisation error for about a factor of four reduced.

Discretisation error by the left-right differences has the magnitude of $(1/\Delta x)$, i.e. a halving of the increment will diskretisation error roughly half. Despite the favorable error the central difference is not suitable for all problems. For convection-diffusion problems, in which the convection stream dominates, it is often too unphysical e.g. oscillations overshoot the numerical solution (Fig. 4.9) when the grid lengths are not small enough. In such cases often preferred unilateral differences, with the flow direction must be taken into account. If the flow in the example shown in Fig. 4.8 pass the grid from west to east as the calculation must be done by the left-hand difference in a reverse flow direction facing with the right difference. Because of the dependence of the flow direction Upwind discretisation method is described. The Upwind discretization of 1st order has the disadvantage that sometimes unrealistic, numerical diffusion effects are introduced (Fig. 4.9). Nevertheless, they occur by their numerical stability in many turbulent flows therefore for the solution of the equations of the k-ε turbulence model is used. There exist Upwind discretizations of higher orders but these are not available in the CFX or FLUENT solver.

96   4 Some important topics of CFD used in this work

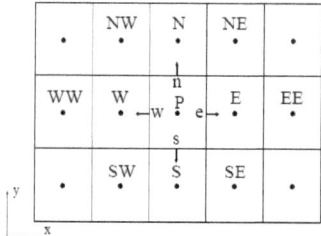

Fig. 4.8: Notation in a staggered two-dimensional numerical grid. The abbreviations N, E, S and W indicate the relative positions north, east, south and west in relation to the point P. In addition, the flows n, e, s and w in the neighboring cells are displayed.

In the context of this work is for the discretization in the three directions of all except the equations of turbulence called high resolution method is used. For the value of Φ_P at the point P is given [83]:

$$\Phi_P = \Phi_{UP} + \beta \nabla \Phi \Delta \vec{v} \tag{4.103}$$

with the value Φ_{UP} Upwind point and vector $\vec{v}$, the vector directed to point P. The factor β lies in the range from zero to one. For values $0 < \beta < 1$ is $\nabla \Phi$ equal to the average of neighboring gradients, in the case $\beta = 1$ is equal to the gradient of Upwind point. Thus determines whether a method first or second order is applied. In areas where the size considered only relatively low gradient, is used as close as possible to 1, approximated in areas of high volatility, however, its tends to zero. In this way the benefits of precision of second order methods will be combined with the numerical stability of the Upwind method.

4.3.1.2 Geometry and mesh generation

Coupled solver and multi-grid procedure

Application of the finite volume method to all elements in the balance field leads to a discrete set of linear balance equations in the form

$$\sum_{nb_i} a_i^{nb} \Phi_i = b_i \tag{4.104}$$

with the number of i control volumes and the knots point P, the solution Φ, the coefficient a of the equation, the right side b and the neighbors nb of i. For a scalar size are all a_i^{nb}, Φ_{nb} and b_i one number, for the coupled set of balance equations for mass and momentum both are a matrix or a vector [83]:

$$a_i^{nb} = \begin{bmatrix} a_{u_x u_x} & a_{u_x u_y} & a_{u_x u_z} & a_{u_x p} \\ a_{u_y u_x} & a_{u_y u_y} & a_{u_y u_z} & a_{u_y p} \\ a_{u_z u_x} & a_{u_z u_y} & a_{u_z u_z} & a_{u_z p} \\ a_{p u_x} & a_{p u_y} & a_{p u_z} & a_{pp} \end{bmatrix}_i^{nb}, \quad \Phi_i = \begin{bmatrix} u_x \\ u_y \\ u_z \\ p \end{bmatrix}_i, \quad b_i = \begin{bmatrix} b_{u_x} \\ b_{u_y} \\ b_{u_z} \\ b_p \end{bmatrix}_i. \quad (4.105)$$

The solver of ANSYS CFX (version 11) uses the coupling of the equations when all rows are using the same solution methods. The advantage of this method, in contrary to the non-coupled or iterative solver used by FLUENT (version 12) lies in its stability and its simplicity [84], and as a disadvantage, higher memory requirements. The overall solution strategy is illustrated in Fig. 4.9. For each time step occur two compute intensive operations. First, the non-linear equations must become linear to get the coefficients for the solution of matrix. Thereafter, the linear equations must be solved by using an algebraic multigrid method. In the stationary case, the iterations are controlled by the time step size to target the solution. In transient calculations the time step size or the maximum number of iterations per time step are performed by the user. In the presented solution strategy is therefore an iterative method in which the exact solution by repeated calculations is approximated. CFX-11 is used to solve the discrete system of linear equations through a multigrid accelerated procedure ILU (incomplete lower upper) solver. The linear system of discrete equations is taking into account the coefficient matrix [A] and the solution vector [Φ] to be formulated as follows [83]:

$$[A][\Phi]=[b]. \quad (4.106)$$

With an iterative method is an approximate solution considered to be improved by a correction factor [83]:

$$\Phi_{n+1} = \Phi_n + \Phi'. \quad (4.107)$$

A corrector value is solution of the equation [83]

$$A\Phi' = r_n, \quad (4.108a)$$

and r_n is residual from $r_n = b - A\Phi_n$. $\quad (4.108b)$

98 4 Some important topics of CFD used in this work

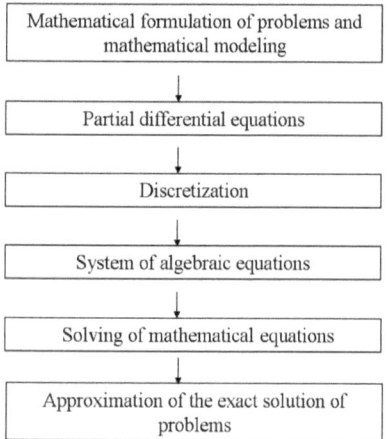

Fig. 4.9: Solution strategy for reactive flow phenomena by using CFD [83].

The computational steps are repeated until the residuals become below a desired limit. Iterative solvers have several disadvantages. Their accuracy decreases rapidly with increasing lengths of the different edges of control volumes. In addition, only errors whose wavelengths are in the range of cells lengths will be quickly filtered. Error larger wavelengths appear very long until they disappear from the solution. This problem can be turned off by the calculations carried out on several grids with different grid lengths. In increasingly roughness of grids appear originally long wave disturbances as relatively small and are quickly filtered out. To avoid the user effort in making a multiple meshing of the geometry, the system of discrete equations for a coarser grid by summation of the equations of the finer grid is created. The result is a gradual, virtual roughness of the grid (Fig. 4.10).

The increased numerical effort by calculating on multiple grids will be faster through the filters of the long wave errors more than offset. By the algebraic multigrid method, the number of iterations required to converge is partially drastically reduced. The computationally intensive process of discretization of the non linear equations needs to be carried out on the finest grid.

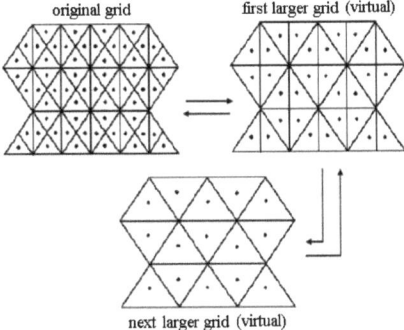

Fig. 4.10: Multigrid procedure: generation of larger grid strikes so as virtual amalgamation of individual cell [83].

4.4 Procedure of CFD simulations

In this work, JP-4 pool fire is simulated as a prototype of sooty hydrocarbon fires. The CFD simulation is done by using different sub-models contained in softwares ANSYS CFX and FLUENT. The different pool diameters are used to show that the experimentally found dependence of various parameters such as temperature and thermal radiation on pool diameter in the calculations realistically reproduced the experimental data. An overview of the simulations and the set of sub-models can be found in Tab. 4.1.

The indication of modeling consists of the initial letters of fuel and the pool diameter in meters. The remarks on the geometry and grid generation, selection and configuration of the sub-models and the definition of initial and boundary conditions will be made in subsequent chapters.

The main purpose of CFD simulation was to determine the temperature T, Surface Emissive Power (SEP) and irradiance E. The simulations are started by using k-ε turbulence model with a buoyancy correction term [83,84] to reach a certain height of the flame which refer to the developing stage of the fire (approx. t = 10 s). After a simulation time of t = 10 s it is assumed that the flame is developed and the further

100 4 Some important topics of CFD used in this work

simulation is continued by using Scale Adaptive Simulation (SAS) [83] or Large Eddy Simulation (LES) [84].

Table 4.1. Overview of the modeling

Geometry	3D cylindrical	3D cylindrical	3D cylindrical	3D cylindrical
Pool diameter	2 m	8 m	16 m	25 m
Turbulence model	k-ε, SAS, LES	k-ε, SAS	k-ε, SAS, LES	k-ε, SAS
Combustion model	Flamelet PDF transport	Flamelet Multistep	Flamelet PDF transport	Flamelet
Soot model	Lindstedt Tesner	Lindstedt Magnussen	Lindstedt Tesner	Lindstedt
Radiation model	Monte Carlo Discrete Ordinate	Monte Carlo	Monte Carlo Discrete Ordinate	Monte Carlo
Time interval	$0 \leq t \leq 10$ s $11\,s \leq t \leq 20$ s	$0 \leq t \leq 10$ s $11\,s \leq t \leq 20$ s	$0 \leq t \leq 10$ s $11\,s \leq t \leq 30$ s	$0 \leq t \leq 10$ s $11\,s \leq t \leq 20$ s

4.4.1 Geometry and meshing

4.4.1.1 Geometry and meshing for the fire in calm condition

In the CFD simulation of large JP-4 pool fires a domain is presented as a 3D cylindrical hexahedral non uniform unstructured mesh (Fig. 4.11). The mesh is very refined at the pool surface and in the inner part of domain with increasing cell dimension as moving to the side boundaries. Dimensions of the geometry and mesh refinement for each pool diameter are listed in the Tab. 4.2.

The fire domain initially contains air under ambient conditions: $T_a = 293$ K and $p_a = 1.013$ bar.

The definition of simplified geometry leads to a significant reduction of the time. The geometries are described in detail.

4.4 Procedure of CFD simulations   101

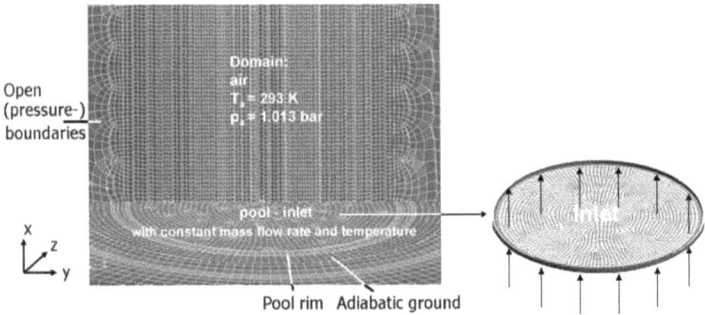

Fig. 4.11: Example of 3D hexahedral unstructured mesh used in simulation of JP-4 pool fires (d = 2 m, 8 m, 16 m and 25 m).

Table 4.2: Dimensions of the geometry and mesh refinement for different d

d (m)	2	8	16	20	25
Dimensions (m^2)					
r · x	3 · 6	10 · 40	20 · 60	30 · 60	30 · 60
Number of cells (10^6)	0.35, 0.8	0.4, 1	1	0.4, 1	1
Cell length (m)					
Min	0.05, 0.01	0.1	0.3	0.30	0.30
Max	0.1, 0.08	0.2	0.3	0.55	0.55
Rim height (m)	0.05	0.1	0.5	0.5	0.5

4.4.1.2 Geometry and meshing for the fire under the wind influence

In a case of CFD simulation of large JP-4 pool fires under the influence of cross-wind coarser meshes are used than in a case of the CFD simulation of a fire in a calm condition. The rectangular geometry is chosen to predict the flame tilt under the wind influence. An example of the geometry is shown on the Fig. 4.12. Dimensions of the geometry and mesh refinement for each pool diameter are listed in the Tab. 4.3.

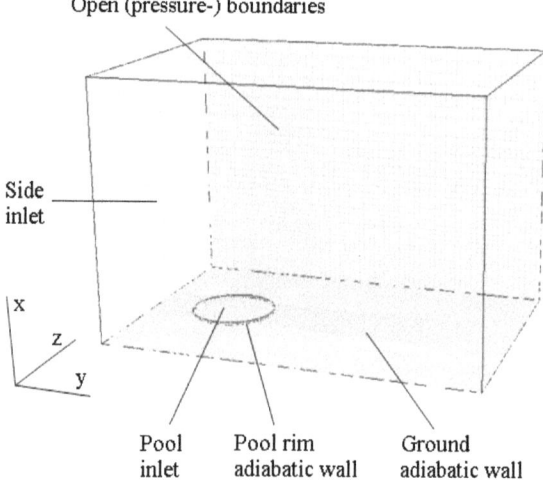

Fig. 4.12: Example of 3D hexahedral unstructured mesh used in simulation JP-4 pool fire (d = 2 m, 20 m and 25 m) under the wind influence.

Table 4.3: Dimensions of the geometry and mesh refinement for different d

d (m)	2	20	25
Dimensions (m^2)			
x · y	6 · 4	60 · 40	60 · 40
Number of cells (10^6)	0.35	0.3	0.3
Cell length (m)			
Min	0.05	0.3	0.3
Max	0.1, 0.08	0.5	0.5
Rim height (m)	0.05	0.1	0.5

4.4.2 Initial and boundary conditions and time steps

The boundaries and the initial conditions are partially identical for two geometries with the difference that an additional inlet boundary surface is used in the case of the geometry used for the simulation of the fire under the wind influence. At that inlet boundary an air flow with a certain velocity which refers to the wind speed is assigned (Fig. 4.12). As the fluid in domain an air is assumed to consist of nitrogen, oxygen and carbon dioxide. The existence of the species carbon dioxide and water at the beginning of the reaction when eddy dissipation model used is essential. The model does not take into account kinetic time scales but the concentrations of various species and the turbulence in the form of k and ε. The mass fractions of individual species in temperature in domain is T_{amb} = 293 K and the pressure is defined relative to the reference pressure p_{amb} = 1.013 bar.

As the mass burning rate experimentally determined value for JP-4 and [25] is used. At the time t = 0 s, the flame is ignited in the experiment, or begins in the computational modeling. From that time initially grows in the size until some time after its full size is reached. In the first test calculations have shown that the flames with a pool diameter 2 m ≤ d ≤ 25 m at the minimum time of t = 10 s reach their full size, the smaller pool flame with d = 2 m already after about t = 5 s.

It appears in the experiments [102,103] that analogous to the growth of the flame, the mass burning rate from a minimum based on a nearly constant value increase. In simulations the constant mass burning rate is used.

For the JP-4 pool fires simulations time steps in a range of 0.0001 s ≤ Δt ≤ 0.01 s are chosen to reach achieved convergence level of minimum 10^{-3} independently on d but depending on the sub-models. Number of iterations per time step varied from 3 to 15 depending on chosen sub-models, where at the beginning of simulation a large number of 10 – 15 iterations per time step is used and after the convergence is reached the number of iterations decreased to minimum 3 – 5. For example in a case of simulations where LES and flamelet model are used, simulation started with the largest time step of 0.01 s and 15 iterations per time step and finally after the achieved convergence is reached a time step of 0.001 s – 0.0025 s is used with 5 iterations per time step, whereas in a case of simulations where SAS and flamelet model are used, simulation started with the smallest time step of 0.0001 s and 10 iterations per time step and when the achieved convergence is reached a time step is increased to the maximal 0.005 s – 0.01 s with 3 – 5 iterations per time step used.

The fuel is assumed to be already evaporated and the fuel vapor coming from the inlet boundary surface with defined constant temperature T_{in} = 373 K and a constant mass flow rate of fuel, $\bar{\dot{m}}_f$. $\bar{\dot{m}}_f$ are based on experimentally determined mass burning rates $\bar{\dot{m}}_f''$ [25]. In Tab. 4.4 $\bar{\dot{m}}_f$ and $\bar{\dot{m}}_f''$ for different d are listed.

The inlet is surrounded with a low rim (Tab. 4.2, 4.3) and adiabatic ground area. The remaining areas in the computational domain are open boundaries with defined ambient conditions. In CFD simulation a non uniform mesh (Tab. 4.2, 4.3) is used with the minimum cell length at the inlet surface and in the near surrounding. With increasing a vertical distance from the pool the cell size increases and reaches the maximum length. The time steps vary depend on the sufficient convergence from $t = 10^{-5}$ s to $t = 10^{-2}$ s. The number of iterations per time step is 5 – 10.

Table 4.4: $\bar{\dot{m}}_f$ and $\bar{\dot{m}}_f''$ of JP-4 pool fire for different d.

d (m)	2	8	16	20	25
$\bar{\dot{m}}_f''$ (kg/(m² s))	0.054	0.054	0.054	0.054	0.054
$\bar{\dot{m}}_f$ (kg/s)	0.17	2.7	10.8	17.0	26.5

JP-4 pool fires (d = 2 m, 20 m and 25 m) were simulated under the influence of cross wind with various wind speed used as corresponding to test data from JP-4 pool fires (d = 2 m, 2.5 m, 3 m, 18.9 m, 20 m and 25 m) [3,9-11,16,18,19]. Wind is defined as an airflow directed to the one of the side boundaries in the rectangular geometry assumed to have a constant velocity at the whole boundary surface. 3D transient simulations are started with zero wind until simulation time of t = 5 s after which a certain wind velocity is applied. After t = 10 s when the relatively steady flame shape is achieved it is assumed that the wind had a certain effect on the flame so that averaging of data is done for a time interval of 10 s ≤ t_b ≤ 20 s. The k-ε turbulence model with buoyancy correction term is used until the t = 10 s of the simulation time after which the LES (d = 2 m) or SAS (d = 20 m and 25 m) is used until t = 20 s. Time steps used were in a range of Δt = 0.00025 – 0.01 s for $0.35 \cdot 10^6$ mesh (d = 2 m) and Δt = 0.005 – 0.02 s for $0.3 \cdot 10^6$ mesh (d = 20 m and 25 m) which minimum is 2.5 times larger than the Δt used in a case of calm conditions for the same mesh refinement.

4.4.3 Determination and configuration of sub-models

The following sub-models are used:
- k-ε turbulence model
- Scale Adaptive Simulation (SAS)
- Large Eddy Simulation (LES) Smagorinsky
- Laminar flamelet model with PDF for non premixed combustion containing 112 species and 800 chemical reaction,
- PDF transport model containing 20 species and 40 chemical reaction,
- Multistep chemical reaction containing 6 species and 6 chemical reaction,
- Monte Carlo radiation model,
- Discrete Ordinate radiation model
- Magnussen soot model,
- Linstedt soot model,
- Tesner soot model.

4.4.3.1 Sub-models for turbulence
4.4.3.1.1 k-ε model

In Section 4.2.1. several turbulence models have already been discussed. The most commonly used two-equation models such as the k-ε model have not initially been developed for certain buoyancy, anisotropic, turbulent natural convection flows. Large eddy simulation appear in the description of such flows very capable, but its demands a large computational effort since the vortex are then calculated directly, without any model assumptions, the model demands a very fine discretization in space and time. This means a fine meshing of geometry, in particular, but also a choice of a very small time step of $t = 10^{-4}$ s is necessary. As the flames in the context of this work are simulated over a period of about $t_{tot} = 20$ s $-$ 30 s, a too high computational effort was in demand when LES and SAS models are used with flamelet models. With the expected increase in computing power of modern computers and the possibility of parallel calculations on inexpensive PC clusters to implement, a standard use of large eddy simulations for pool flames in the near future can be expected. The k-ε model with buoyancy terms needed, in comparison to LES, a much larger time steps of 0.01 s $\leq \Delta t \leq$ 0.05 s, a significantly less computationally intensive, and also can give fairly good results with in modeling similar flows [17-19].

106   4 Some important topics of CFD used in this work

In this work the CFD simulation is started at $t_0 = 0$ with a k-ε turbulence model with a buoyancy correction term [83,84] to reach a certain height of the flame. At t = 10 s the flame is developed and for t > 10 s during a burning time of t_b = 10 s a further simulation of a flame is performed by using Large Eddy Simulation (LES) or Scale Adaptive Simulation (SAS). The burning time t_b starts at 10 s when the flame is developed and ends at 20 s limited by the CPU time. This means that the flame during the burning time of 10 s show a real burning.

4.4.3.1.2 Large Eddy Simulations (LES) Smagorinsky

Here, dynamic LES Smagorinsky model is used in simulation with FLUENT software.

The Large Eddy Simulation (LES) is a method for numerical calculation of turbulent flows with high Reynolds numbers. These are the Navier-Stokes equations locally and coincided with a low pass filter filtered. Thus, the large vortex structures are directly calculated and the small structures are modeled with one of the RANS turbulence models [104]. The computing lies between direct numerical simulation (DNS) and the solution of Reynolds equations, RANS (Reynolds averaged Navier Stokes) which only calculates an averaged value, so the computation time is significantly lower than in the DNS.

4.4.3.1.3 Scale Adaptive Simulations

In this work SAS is used in simulation with CFX software.

Scale Adaptive Simulation (SAS) models are advanced URANS (Unsteady RANS) models which allows a realistic calculation of unsteady flows. Unlike standard URANS methods that typically resolve only large scale structures SAS allows a decline of greater turbulence elements in a turbulent spectrum. SAS achieved the improvement in the transient field by taking into account the Karman length scale in the scaling of the equation turbulence model. This allows the model already resolved structures and avoids the smearing which is in the standard URANS methods typical. A detailed mathematical description can be found in CFX guideline [83].

4.4.3.2 Sub-models for combustion

For the modeling of the JP-4 pool fires, the Eddy dissipation model and the different flamelet models are chosen. The choice of combustion model is done depending on simulation type. For the simulation of fires in a calm condition, mostly a laminar flamelet model has been used. Some simulations are done by using different combustion models e.g. flamelet model and Eddy dissipation model e.g. JP-4 pool fire with d = 8 m and d = 25 m to investigate the influence of combustion models on predicted temperature and SEP of the fire. The simulations with the fires under the wind influence are done in some cases (d = 20 m and 25 m) by using Eddy dissipation model with multistep chemical reaction (6 reactions and 6 species).

The choice of the combustion model, depending on the grid refinement and environment fire condition simulated.

4.4.3.3 Sub-models for thermal radiation

While the heat conduction in solids is the only type of heat transport, in turbulent flows plays a minor role. Here dominate depending on the temperature, either the convection or thermal radiation. The different modeling of the thermal radiation in the simulation program ANSYS CFX and FLUENT can be found in [83,84].

The balance equation for the thermal radiation is given as

$$\frac{dL_\lambda(r,\vec{s})}{dl} = \underbrace{-(k_{abs,\lambda} + k_{scat,\lambda})L_\lambda(r,\vec{s})}_{(a)} + \underbrace{k_{abs,\lambda}L_\lambda(r,\vec{s},T)}_{(b)} + \underbrace{k_{scat,\lambda}\int dL_\lambda(r,s)\psi(\vec{s},\vec{s}',\Omega)d\Omega}_{(c)} + \underbrace{S_{L\lambda}}_{(d)} \qquad (4.109)$$

Term (a) with the absorption k_{abs} and scattering coefficients and k_{scat} refer to increasing the radiation intensity in the direction r,s; term (b) refer to the black body radiation with k_{abs}; term (c) refer to the re-emitted radiation from all other directions with k_{scat} and term (d) refer to the additional heat source $S_{L\lambda}$.

In the following text the different radiation models used in this work are discussed.

4.4.3.3.1 Discrete Ordinate radiation model

In addition to Monte Carlo simulations are also the discrete transfer radiation method in the modeling in pool flames enforced. Here, the radiance L is not through the pursuit of a very high number of photons identified, but along the less number of rays calculated. This way, the computing time compared to the Monte Carlo method significantly shortened. These simplifications, however, are necessary, which may distort the result. The discrete ordinate method [84] assumes an isotropic scattering from. Eq. (4.109) thus simplifies to

$$\frac{dL_\lambda(r, \vec{s})}{dl} = -(k_{abs,\lambda} + k_{scat,\lambda})L_\lambda(r, \vec{s}) + k_{abs,\lambda} L_\lambda(r, \vec{s}, T)$$

$$+ \frac{k_{scat,\lambda}}{4\pi} \int_{\Omega=4\pi} dL_\lambda(r, \vec{s}')d\Omega + S_{L_\lambda}$$

(4.110)

Furthermore it is assumed that the balance area has relatively homogeneous radiation characteristics, such that [83]:

$$L_\lambda(r) \approx L_\lambda(r + dr) \tag{4.111}$$

The intensity is then along rays, with L, 0 from the edges of the balance area, using the following equation [95]:

$$L_\lambda(r,\vec{s}) = L_{\lambda,0} \exp(-(k_{abs,\lambda} + k_{scat,\lambda})l) + L_{S,\lambda}(1 - \exp(-k_{abs,\lambda}l)) + k_{scat,\lambda}\overline{L_\lambda}. \tag{4.112}$$

Usually, 8 – 16 rays in the balance area are prosecuted. The extension of the results along the ray paths to the entire record area is under the above assumption of homogeneity of the area. In the case of a gray medium, the intensities are only calculated once per ray. If, however, for example, different bands of a gas have to be taken into account, the first solution for each corresponding wavelength can be calculated. Finally, the integration over the individual results is done to determine the overall intensity.

4.4.3.3.2 Monte Carlo radiation model

Monte Carlo simulations are applicable for many different problems, such as in meteorology to forecast climate change, or even in the financial industry for the prediction of exchange rates. The Monte Carlo method is one of the most robust numerical methods to calculate the heat transfer by radiation. Its advantage lies in the possibility to simulate the radiation heat transfer in arbitrary geometric configurations, taking into account the spatially varying optical properties of a medium radiation in a simple way. When used the Monte Carlo method is a representative number of photons in the entire simulation area emitted. The starting points and the beginning of the individual photons are randomly chosen. Therefore, it must be ensured that a sufficiently large number of photons is created to ensure a realistic picture of radiation. The probability that a photon in a particular place, or in a given cell along the beam path is scattered or absorbed, is determined by the spatially variable absorption k_{abs} or k_{scat} scattering coefficients. In the following chapters k_{abs} will be replaced with a modified absorption coefficient $\bar{\bar{æ}}$. The irradiance E in a given area is equal to the number of counted on it taken photons multiplied by the associated emissivity [83]. The crucial disadvantage of the Monte Carlo method is that it is very computationally intensive, if a good accuracy of results is desired. This is particularly true when the Monte Carlo method as in the context of this work with other methods such as finite differences or finite volume method is coupled. Due to the large number of cells, the number of required computational steps is accordingly high. Also, the stability of time integration schemes in the case of transient problems is heavily influenced by the accuracy of the results of a Monte Carlo simulation. The spatially and temporally varying temperatures and species concentrations in flames cause the equally variable emission or absorption.

4.4.3.4 Sub-models for soot
4.3.3.1 Magnussen soot model

The soot modeling according to Magnussen [92] represents a direct extension of the Eddy dissipation model for the calculation of combustion products such as soot and NO_x. For calculation of the soot in the Magnussen model several steps are used:

Source terms of species

    1. Soot formation in the form of radicals,

    2. Formation of soot particles from the radicals,

    3. Oxidation of soot particles and radicals,

    4. Calculation of the influence of turbulence on the reaction rates with Magnussen and Eddy dissipation model.

The two sizes will be required: for the soot mass fraction Y_s and the specific concentration of radicals X_N. The two balance equations contain a term for formation and oxidation the respective sizes. More detail description of the model can be found in Section 4.2.4.1.

Table 4.5: Standard values of parameters in Magnussen model [83]

Parameter	Value	Value (mol)
ρ_s	2000 [kg/m^3]	
d_s	1.75 · 10^{-8} [m]	
a_0	1.35 · 10^{37} [1/(kg s)]	2.24 · 10^{13} [mol/(kg s)]
f_C	carbon mass fraction in fuel (e.g. 12/16 for methane)	
$T_{A,0}$	90000 [K]	
f-g	100 [1/s]	
g_0	1.0 · 10^{-15} [m^3/ s]	6.02 · 10^8 [m^3/(s mol)]
a	1.0 · 10^5 [1/s]	
b	8.0 · 10^{-14} [m^3/ s]	4.98 · 10^{10} [m^3/(s mol)]

4.3.3.2 Soot modeling with Lindstedt

Lindstedt model leads to soot modeling two sizes, for one so-called particle density N, the number of particles per mass of mixture states, and continue to soot mass fraction Y_s [83,105]. The determination of the source terms S_N and S_{Y_s} is described in Section 4.2.4.2. Experiments show that acetylene plays an important role in the different processes of soot occupies (Chap. 2.5.3). For the first acetylene molecule PAHs can be built later in the HACA mechanism which can be involved in sub-models. Accordingly, Lindstedt describes the nucleation and the growth of surface as a function of acetylene concentration. Details in modeling of soot formation can be found in [91,105].

4.3.3.3 Tesner soot model

The formation of radicals and soot are modeled according to Tesner [93]. The formation of soot and nuclei particles is given as [84,93]:

The equations included in the model depend on the formation and oxidation of soot particles and the radicals from the fuel mass fraction from. This fact represents a large numerical advantage, for the source of the JP-4 as a fuel calculated with the Eddy dissipation model and represents the influence of turbulence on combustion. As the source of the solution according to the fuel balance equation exists, no additional computing time steps for the consideration of turbulence on soot are required. This direct dependence can also prove detrimental. Since the fuel mass fraction is calculated here in the large pool fire in a relatively low dimensionless heights x/d mainly with increasing distance from the flame axis reach very small values, the oxidation terms for the soot particles and the radicals in many areas of the flame is also very small. Considering the relatively complicated soot formation mechanism consists of on the various intermediate steps such as nucleation, formation of the primary particles, particle growth and oxidation, in the soot oxidation model a significant simplification is used in CFD models. More details about the model can be found in Section 4.2.4.3.

4.4.4 Modeling of absorption coefficient of the flame

Absorption coefficients used in the radiation model OSRAMO II, III [7,27] contained in effective absorption coefficient of the flame $\bar{\bar{æ}}_{eff}(T)$ assume influence of air absorption $\bar{\bar{æ}}_{air}(T)$ = 0.02 m^{-1}, absorption of soot parcels $\bar{\bar{æ}}_{sp}(T)$ = 1.035 m^{-1}, hot spots $\bar{\bar{æ}}_{hs}(T)$ = 0.404 m^{-1} and reactive zone $\bar{\bar{æ}}_{re}(T)$ = 0.380 m^{-1}. Sinai et al [17] use step function of averaged absorption coefficient $\bar{\bar{æ}}$ which contains absorption coefficient of air $\bar{\bar{æ}}_{air}$ = 0.02 m^{-1} and $\bar{\bar{æ}}$ of fuel/air reacting mixture $\bar{\bar{æ}}_{f+air}$ = 0.5 m^{-1} which decrease the influence of soot absorption. Mc Grattan [15] use constant $\bar{\bar{æ}}$ = 0.05 m^{-1} of a flame (gaseous combustion mixture of fuel and air) included in SFM, neglecting a smoke blockage effect which lead to a noticeable increase of calculated surface emissive power. In [25] a constant absorption coefficient $\bar{\bar{æ}}$ = 0.5 m^{-1} is given for JP-5 gaseous fuel and $\bar{\bar{æ}}$ = 2.6 m^{-1} for liquid kerosene fuel.

Tien et. al obtain the following expression for the soot absorption-emission coefficient by integration of the soot emission over all wavelengths:

$$\bar{\bar{æ}} = 3.6\ B_R f_V T/C_2 = C_{K,R} f_V T \qquad (4.113)$$

where B_R is dimensionless soot constant, $f_V = \dfrac{\rho_2 Y_2}{\rho_s}$ is a volume fraction of soot and C_2 is Planck´s second constant.

Lautenberger et.al use the Eq. 4.113 with the following values: $C_{K,R} = 1226$ (m/K) and $B_R = 4.9$. In [107] absorption coefficient $\bar{æ}$ is given as the sum of gas-phase and soot contributions $\bar{æ} = \bar{æ}_g + \bar{æ}_s$. The gas absorption coefficient $\bar{æ}_g$ is evaluated in terms of the local temperature and mixture fraction using the narrow band radiation model, RADCAL [107]. The spectral absorption-emission coefficient, $\bar{æ}_{s\lambda}$ of soot in the small particle limit is proportional to the soot volume fraction divided by the wavelength λ:

$$\bar{æ}_s = \frac{B_s f_V}{\lambda} \qquad (4.114)$$

where B_s is a dimensionless constant based on the soot complex index of refraction [107].

Jensen et. al determines the absorption coefficient of fire based on the contributions mostly from soot particles but also from carbon dioxide and water vapor absorption at selected wavelengths with an empirically based wideband model. The model for these coefficients assumes that the medium is gray and the soot is dominant absorbing and emitting species and it is defined as the broadband emitter the spectral absorption based on the spectral law on the whole frequency spectrum in order to obtain Planck mean coefficient which is used in radiation transport equation (RTE) [29].

5. Results and discussions
5.1 Instantaneous and time averaged flame temperatures
5.1.1 Thermograms

The local distribution of temperatures at the flame surface, defined as emission temperatures T(x, y) is obtained experimentally directly from a thermogram (Figs. 5.1a1 and 5.2) which received thermal radiation from the fire [3,4]. For the emissivity of large, sooty fire $\varepsilon_F = 0.92$ is used [2,3]. The IR images obtained from the thermographic camera were transferred as matrices of temperature using the software provided with the thermographic camera [3,4]. Each instantaneous thermogram, consists of the pixel elements of a matrix with 480 rows and 640 columns where $T_{i,j}$ is emission temperature of a pixel element at the position (i, j) in a x, y plane of the thermogram [3,4,27]. The instantaneous thermogram consists of colored areas which correspond to a defined temperature range ΔT or to a defined range of surface emissive power ΔSEP [3,4,27]. A series of thermograms of a JP-4 pool fire (d = 16 m) are presented in Fig. 5.1a1 and four instantaneous thermograms of a JP-4 pool fire (d = 25 m) are shown in Fig. 5.2.

The instantaneous thermograms of JP-4 fires (Fig. 5.1a1 and 5.2) show local inhomogeneities of temperature and time dependent fluctuations in these flames. The time averaged emission temperatures $\overline{T}_{i,j}$ of pixel elements are obtained from a series of thermograms by using the instantaneous temperatures $T_{i,j}$ [3,27]:

$$\overline{T}_{i,j} = \frac{\sum_{1}^{N_T} T_{i,j}}{N_T} \qquad (5.1)$$

By CFD simulations the emission temperatures are calculated from SEP by using Stephan Boltzman law and the SEP is obtained by integration of incident radiation G along the path length. Detailed procedure is given in Chapter 5.2.6.

The procedure is done as follows:
- The flame is divided into large number (e.g. 50) of parallel planes perpendicular to the pool surface,
- On the each plane a distribution of incident radiation G(x, y) is plotted,
- The certain path length s between $d/4 \leq s \leq d/2$ through the parallel planes starting from the flame centerline is chosen for integration of G,

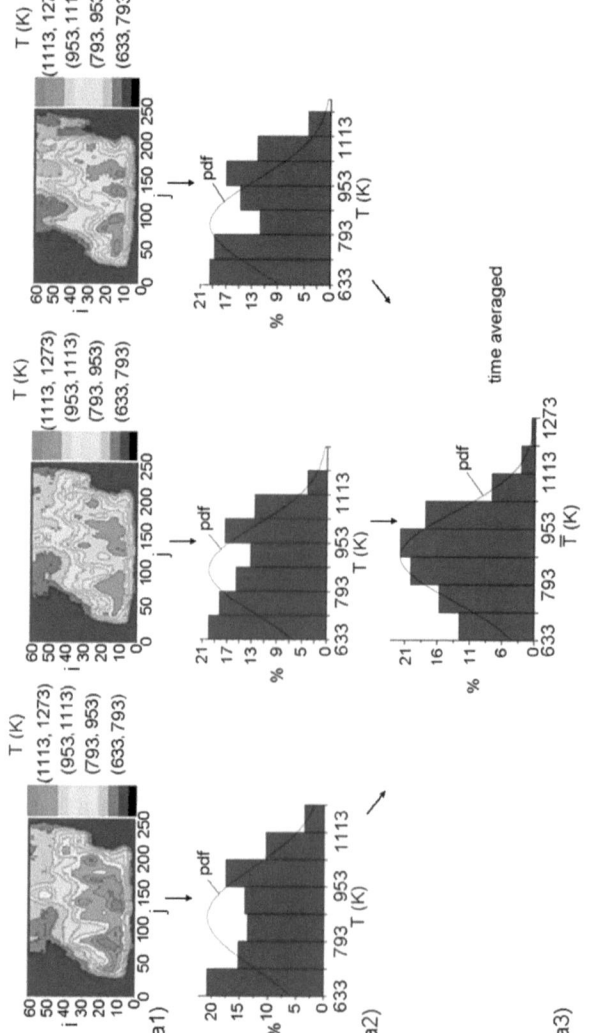

Fig. 5.1a: Evaluation of instantaneous histograms h(T) (a2) and time averaged histogram $\bar{h}(T)$ (a3) of temperature from thermograms (a1) of JP-4 pool fire (d = 16 m).

5.1 Instantaneous and time averaged flame temperatures 115

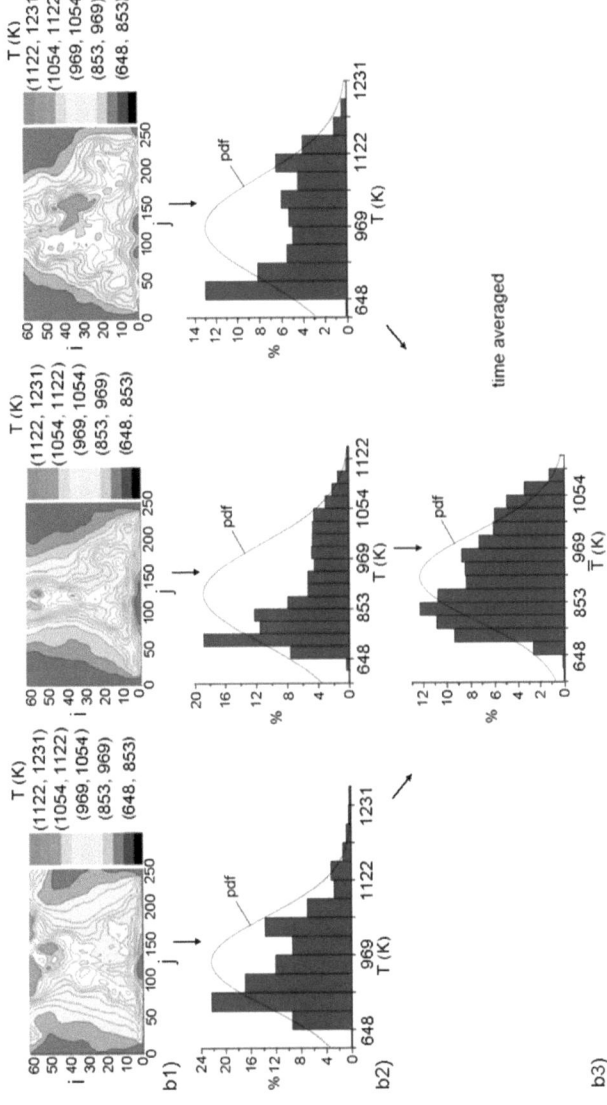

Fig. 5.1b: CFD predicted instantaneous histograms $h_{CFD}(T)$ (b2) and time averaged histogram $\overline{h}_{CFD}(T)$ (b3) of temperature of JP-4 pool fire (d = 16 m).

116 5. Results and discussions

- The distribution of G on the parallel planes is integrated along the path length s which results with an integrated distribution of G(x,y) given on the resulting plane. The incident radiation in each point in the resulting plane G(x,y) is then transferred by the Stephan Boltzman law into temperature T(x,y).

-The integration process is done for each time step so the instantaneous temperature distributions are obtained. The CFD predicted instantaneous temperature distribution T(x,y) (Fig. 5.1b1) can be assigned as temperatures of pixel elements $T_{i,j}(t)$ to compare with thermograms (Fig. 5.1a1).

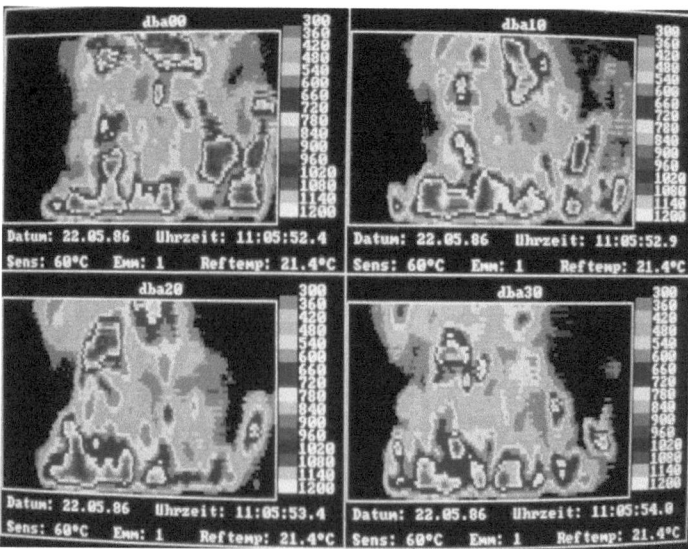

Fig. 5.2: Instantaneous thermograms of JP-4 pool fire (d = 25 m).

5.1.2 Histograms

The instantaneous temperature distributions of a fire can be presented by histograms h(T) (Fig. 5.1a2) determined from instantaneous thermograms (Fig. 5.1a1) showing frequency of temperature distribution in the fire.

The instantaneous histograms h(T) of temperature (Fig. 5.1a2) of a JP-4 fire (d = 16 m) show a local inhomogeneities and time dependent fluctuations in the flame. The time averaged histogram $\overline{h}(T)$ (Fig. 1a3) obtained by averaging the instantaneous histograms with the number N_T (e.g. 50) shows a large frequency of emission temperatures in interval of 793 K $\leq \overline{T}_{exp} \leq$ 1033 K [3,4]. Low emission temperatures are typical for large sooty hydrocarbon pool fires where smoke blockage effect plays a great role in decreasing of thermal radiation to the surrounding and hence reduction of emission temperatures. In JP-4 pool fire a relatively large amount of soot is produced which leads to an increase of thermal radiation of a flame and therefore to an increase of the flame temperature [27]. However, this increase of the emission temperature is partially compensated by the smoke blockage effect occurring in the most of hydrocarbon pool fires, which leads to a certain decrease of thermal radiation and therefore to a decrease of the emission temperature [27].

The CFD predicted instantaneous histograms $h_{CFD}(T)$ (Fig. 5.1b2) obtained from thermograms (Fig. 5.1b1) are averaged over a real burning time t_b = 10 s to get time averaged histogram $\overline{h}_{CFD}(T)$ as shown in the Fig. 5.1b3. The histogram shows a large frequency of emission temperatures in interval of 700 K $\leq \overline{T}_{CFD} \leq$ 904 K. A discrepancy between CFD predicted results and the measured data are due to the path length s chosen for the integration of G. The path length s in ideal conditions should mimic real path length of the JP-4 pool fire (d = 16 m) what is not easy to determine in time dependent CFD predicted results, neither in experiments.

5.1.3 Probability density function (pdf)

The time averaged pdf of temperature is determined based on the time averaged histogram (Fig. 5.1a3). The time averaged temperature at the maximum of the pdf predicted by CFD, $\overline{T}_{CFD}$ = 911 K (Fig. 1b3) is for about 38 K higher than the time averaged measured temperature $\overline{T}$ = 873 K at the maximum of pdf (Fig. 5.1a3) [3,4]. It must be noticed that the thermograms used here for comparison of CFD results belongs to one of the series of thermograms of JP-4 pool fire (d = 16 m) given in [3,4] and the time averaged temperature from thermograms vary from test to test. The temperatures $\overline{T}^*$ obtained by Gaussian distribution from the measured $\overline{T}$ (Table 5.1) [3,4] can be used for comparison with $\overline{T}_{CFD}$. It can be seen that the $\overline{T}_{CFD}$ = 911 K is only slightly larger than the averaged temperature from series of experiments $\overline{T}_{exp}$ = 902 K [3,4].

5. Results and discussions

Table 5.1: Averaged values and standard deviation for different thermographic series for JP-4 pool fire (d = 16 m)

Serie	$(x_* = \bar{x})^1$	s	$\overline{T^*}$ (K)	$\overline{T}_{exp}$ (K)	$\overline{T}_{CFD}$ (K)
A	– 0.1651	0.175	848		
B	– 0.1001	0.200	905	902	911
C	– 0.0896	0.188	914		
D	– 0.1290	0.168	879		

[1] based on Gauss normal distribution

Table 5.2: Averaged values and standard deviation for different thermographic series for JP-4 pool fire with different d

d(m)	$(x_* = \bar{x})^1$	s	$\overline{T}_{exp}$ (K)	$\overline{T}_{CFD}$ (K)
8	– 0.043	0.190	975	980
16	– 0.1205	0.184	902	911
25	– 0.1169	0.187	859	900

[1] based on Gauss normal distribution

$$\int_{x=-\infty}^{x_*} G_x(x)dx = \frac{1}{2} \qquad (5.2a)$$

where

x_* is a mean value

$$\overline{T^*} = T_0 \cdot e^{x_*} = T_0 \cdot e^x \qquad (5.2b)$$

$x = \ln(T/T_0)$ with $T_0 = 1000$ K (5.2c)

Due to the limited number of series of instantaneous thermograms in a visual form, only one serie of $N_T = 50$ pictures is used for visual comparison with the CFD results. Also, it should be noted that the slight wind with velocity of changeable strength and direction was present during the experiments [3,4] which could have an influence on the heat flux received by thermograms and a temperature distribution, hence the discrepancy between the measured and CFD results, here in $\overline{T}_{exp}$ and $\overline{T}_{CFD}$. Another reason is due to the path length s used for integration of G by CFD (Chapter 5.2.6).

Detailed calculation can be found in [3]. Validation of CFD simulation by using global metric [107] is given in Chapter 5.5.

5.1.4 Temperature fields

A quantitative description of the dynamics of JP-4 pool fires (d = 2 m, 8 m, 16 m and 25 m) is shown in Fig. 5.3.1a,b-5.3.4a,b by the predicted isotherms at three different times. The first instantaneous temperature field applies to the time t_1 = 12 s after ignition, t_2 = 14 s and t_3 = 16 s. In the three fields the flame pulsation is noticeable. The flame pulsation is connected with a formation and rising of vortices which greatly effect the distribution of the maximum flame temperature. A significant expansion of small vortices at the pool rim is visible in all predicted temperature fields. Here, inside the vortices the temperatures are significantly higher than in a surrounding area, for 1200 K ≤ T ≤ 1800 K.

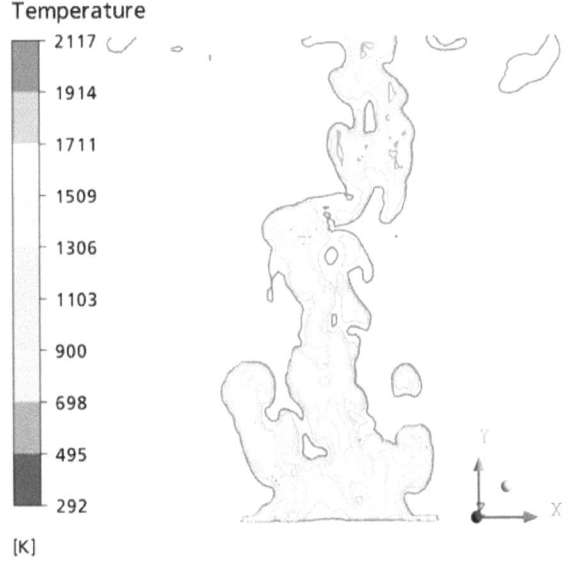

Fig. 5.3.1a: Isotherms of JP-4 pool fire (d = 2 m) at t = 12 s predicted by using flamelet model.

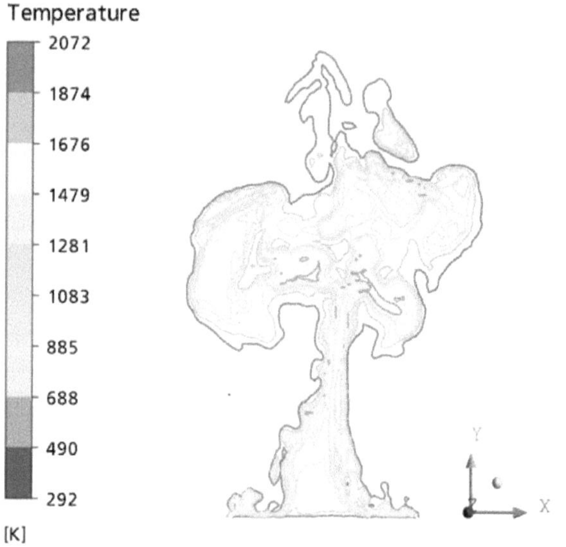

120   5. Results and discussions

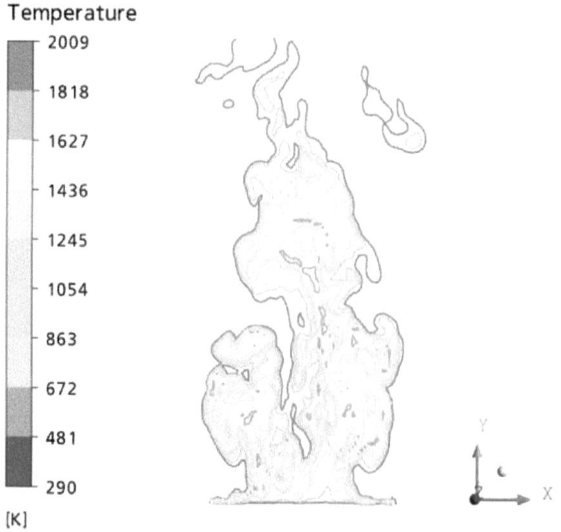

Fig. 5.3.1b: Isotherms of JP-4 pool fire (d = 2 m) at t = 14 s and 16 s predicted by using flamelet model.

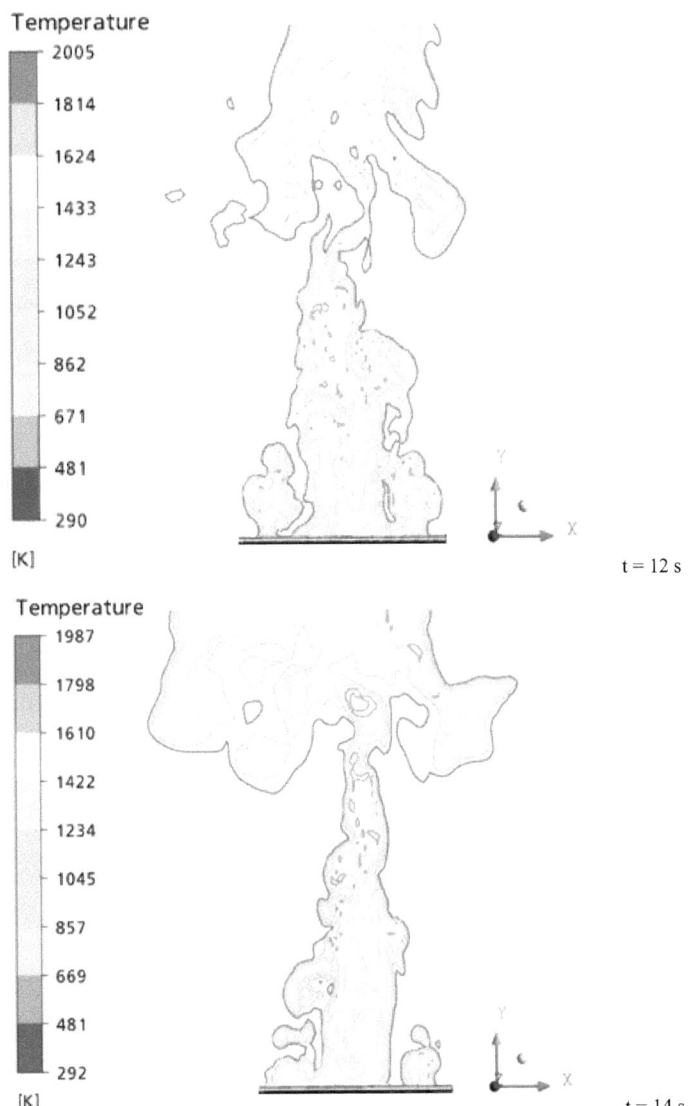

Fig. 5.3.2a: Isotherms of JP-4 pool fire (d = 8 m) at t = 12 s and 14 s predicted by using flamelet model.

122   5. Results and discussions

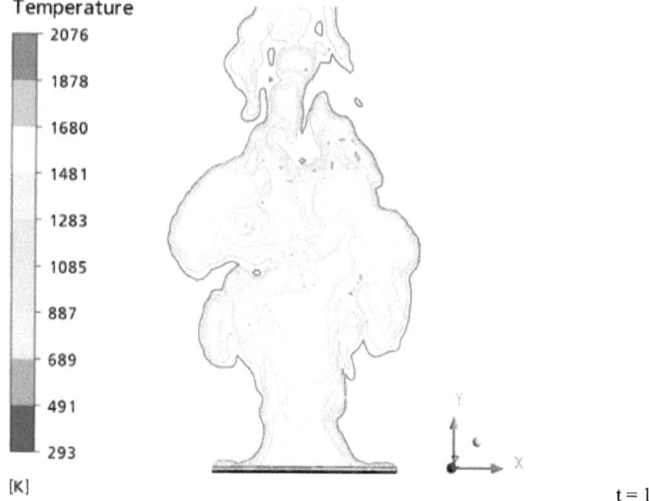

Fig. 5.3.2b: Isotherms of JP-4 pool fire (d = 8 m) at t = 16 s predicted by using flamelet model.

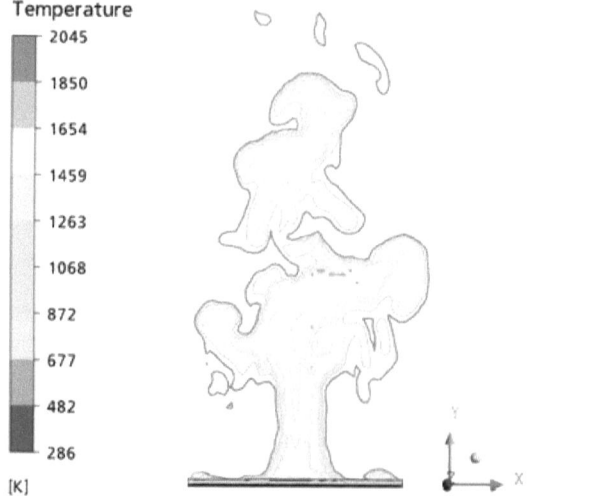

Fig. 5.3.3a: Isotherms of JP-4 pool fire (d = 16 m) at t = 12 s predicted by using flamelet model.

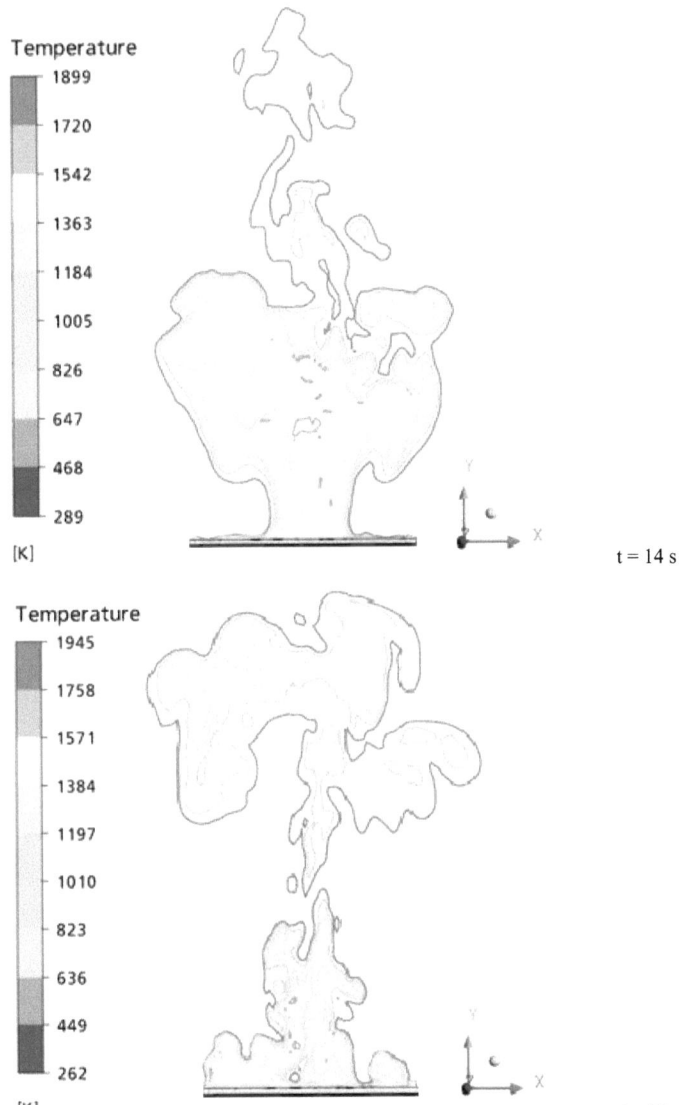

Fig. 5.3.3b: Isotherms of JP-4 pool fire (d = 16 m) at t = 14 s and 16 s predicted by using flamelet model.

124   5. Results and discussions

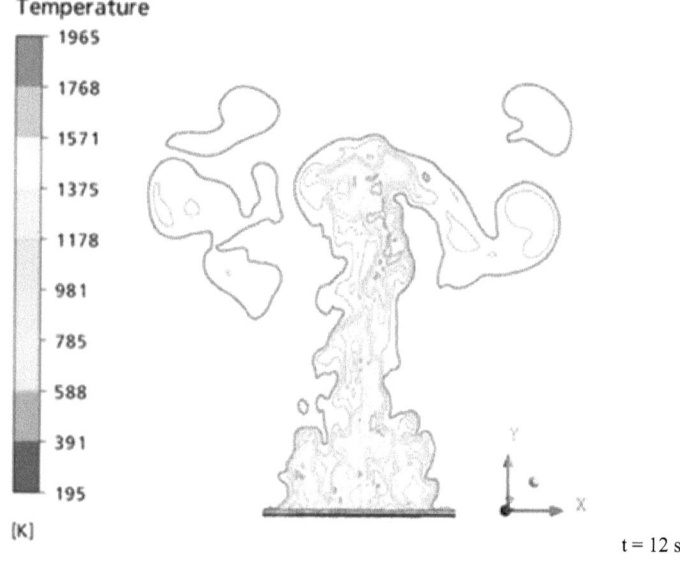

t = 12 s

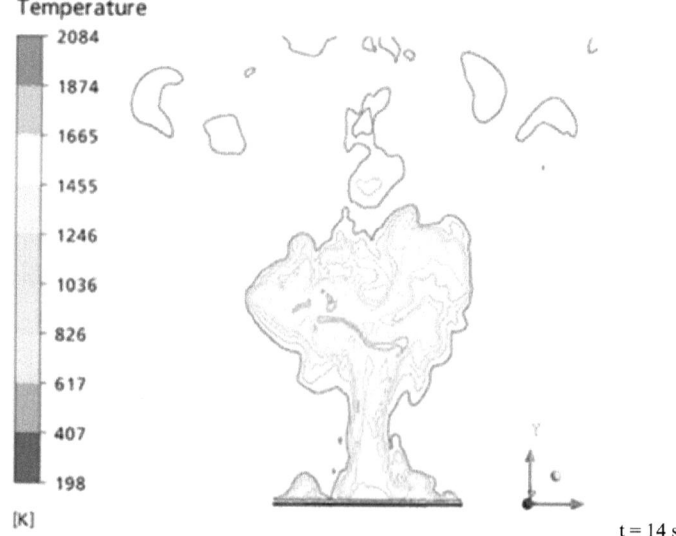

t = 14 s

Fig. 5.3.4a: Isotherms of JP-4 pool fire (d = 25 m) at t = 12 s and 14 s predicted by using flamelet model.

5.1 Instantaneous and time averaged flame temperatures 125

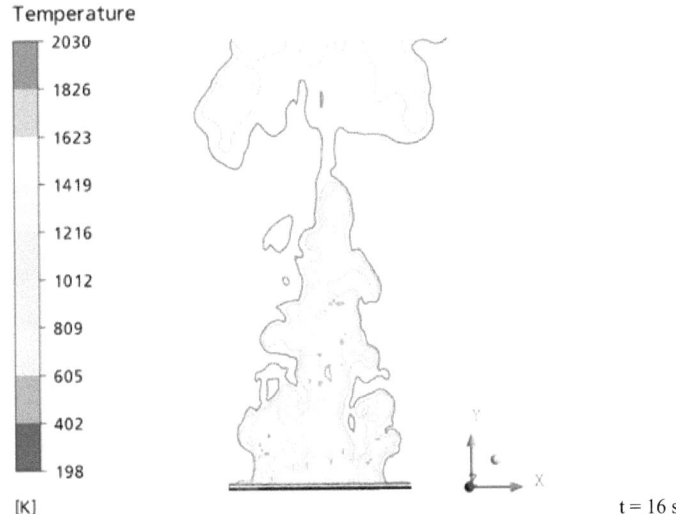

Fig. 5.3.4b: Isotherms of JP-4 pool fire (d = 25 m) at t = 16 s predicted by using flamelet model.

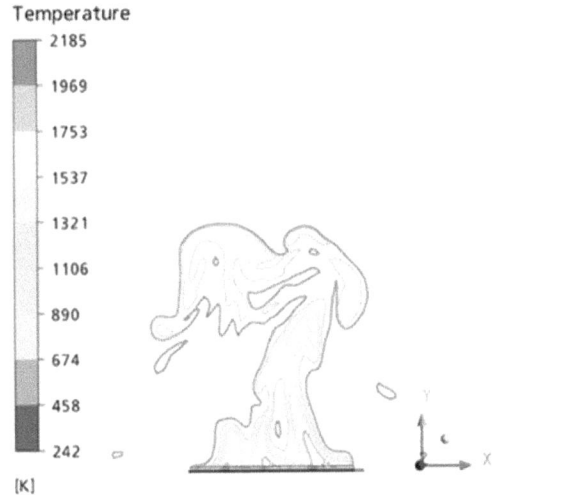

Fig. 5.3.5a: Isotherms of JP-4 pool fire (d = 25 m) at t = 12 s predicted by using multistep chemical reaction.

126   5. Results and discussions

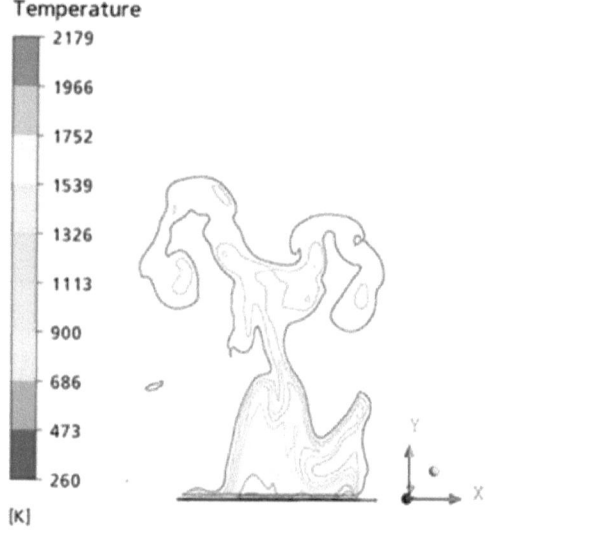

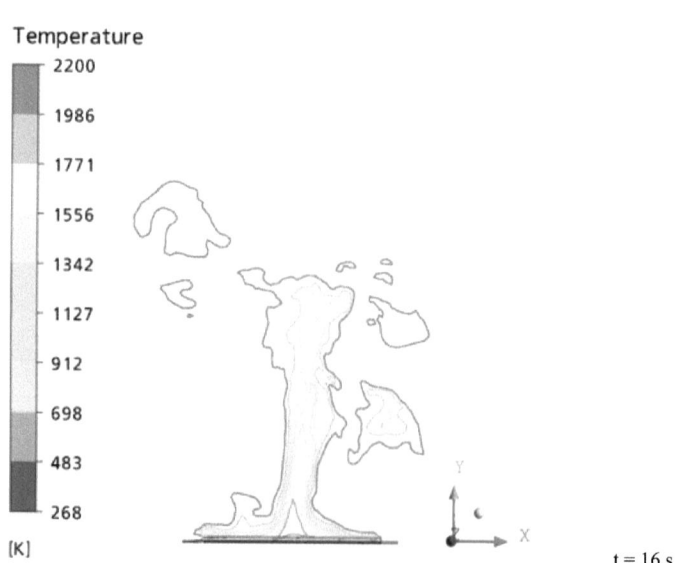

Fig. 5.3.5b: Isotherms of JP-4 pool fire (d = 25 m) at t = 14 s and 16 s predicted by using multistep chemical reaction.

The hottest areas in large pool flames are concentrated inside of vortices at some distance from the pool. The hot flame gases have a significantly lower density (Chapter 2.4), and therefore rise up with increasing time. By rising up they are cooling in the outer area of flame, so that in the upper part of the flame the temperatures of T ≤ 1300 K occur (Fig. 5.3.1a,b-5.3.4a,b).

Analyzing the instantaneous results it can be seen that in a case of JP-4 pool fire with d = 2 m (Fig. 5.3.1a,b) the maximum temperatures are visible also in a upper part of the flame (due to the more frequent occurrence of hot spots) more than in a case of JP-4 fire with larger pool diameter (d = 16 m and 25 m) (Fig. 5.3.3a,b and 5.3.4a,b) where the maximum temperatures are concentrated near the pool and inside of lower vortices. The pulsation of the flames occurs in all simulations, although with different frequencies. This observation is consistent with the experimental results [3]. The temporal fluctuations of the temperature field, have specific effects on the SEP of the flame and thus on the irradiance on neighboring objects and people (Chap. 5.2 and 5.3). In the presentation of the JP-4 flame (Fig. 5.3.1a,b-5.3.4a,b) a typical constriction of the flame on the clear combustion zone and smoky zone is noticeable, especially in a case of larger pool diameters (d = 16 m and 25 m) (Chapter 2.5). Up to the level which separate combustion zone from the upper smoky part, very high temperatures are found. Going to the upper part of the flame temperature significantly decreases. The high temperature occurs mainly in the vicinity of the flame axis at some distance from the pool and also inside the vortices which spread in the radial direction from the flame axis. The presented temperature fields in Fig. 5.3.1a,b-5.3.4a,b are in line with experimental observations [3]. In addition, by comparison of Fig. 5.3.4a,b and 5.3.5a,b, the influence of chemical reaction models on predicted temperature field can be seen. In a case of simplified chemistry used in CFD simulation (multistep chemical reaction) the averaging process is more pronounced and the maximum temperatures are always concentrated in a lower part of the flame (Fig. 5.3.5a,b) and in a case of detailed chemical reaction modeling the predicted temperature field consists of more detailed structures where the hot spots are greatly involved in a broad range of colder flame regions at the lower and upper part of the flame (Fig. 5.3.4a,b).

A closer comparison with measurements is made by using averaged sizes in Chapter 5.1.5.

5.1.5 Axial and radial profiles

In the simulations, as well as in the experiments it is found that the maximum of the time averaged temperatures are along the flame axis away from the centerline. In the simulation, in contrast to most experiments [3,4,9], free flows of external influences such as wind, are avoided. That explains main deviations from experiments [3,4,9]. Due to the time and memory consuming, transient simulations which include detailed chemistry (flamelet model) and turbulence (LES, SAS) models an averaging time for prediction of CFD results is often restricted to the few seconds (e.g. up to 10 s) of simulated time and contains mainly earlier burning time. With an increasing averaging interval the deviation decreases (e.g. from Δt = 5 s to 10 s).

In a relative height x/d = 0.25 is the maximum average axial temperature in the JP-4 pool fire with larger pool diameter $\overline{T}_{max}$ (d = 25 m) = 1230 K (Fig. 5.4.4), however, about 70 K lower than axial $\overline{T}_{max}$ at x/d = 0.65 for the smaller JP-4 pool fire (d = 2 m) = 1300 K (Fig. 5.4.1). The main difference from axial $\overline{T}_{max}$ at relative radial distance y/d = 0 from $\overline{T}_{max}$ at y/d = 0.05 in a case of JP-4 pool fire (d = 25 m) is 30 K.

On the Fig. 5.4.4 it can be seen that the maximum of axial $\overline{T}$ for JP-4 (d = 25 m) is found for y/d = 0.05 and it is moving close to the pool with increasing radial distance from the flame axis y/d. The same situation is found for JP-4 pool fires with smaller d (Fig. 5.4.1-5.4.3) but the discrepancy in axial $\overline{T}_{max}$ depending on y/d decreases with decreasing d. The main difference from the maximum axial temperature at y/d = 0 is about $\overline{T}_{max}$ (d = 16 m) = 10 K at y/d = 0.05 (Fig. 5.4.3).

The bulges in the profile of axial $\overline{T}$ of the large flame at different y/d are caused by hot and slightly to the side rising vortex. The small irregularities in the slope of the curves are due to the averaging period. The maximum temperature is, however, not directly at the flame axis. This may be due to the short averaging time, or even the fact that at low altitudes flame fluctuations are very high.

Because of the very long computational time a substantial extension of the calculations is not practical.

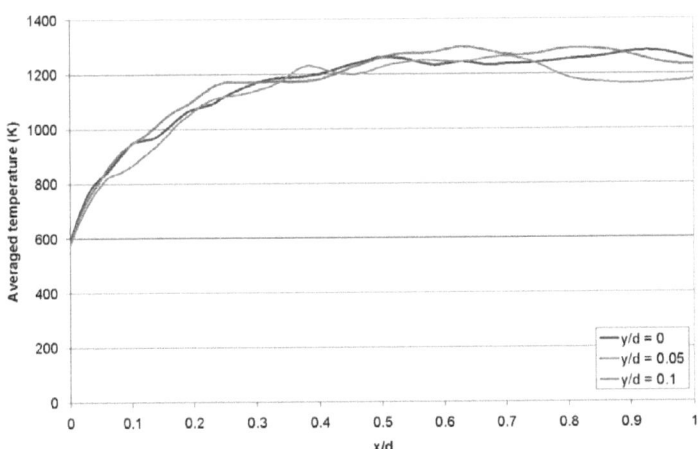

Fig. 5.4.1: Axial time averaged $\overline{T}_{CFD}$ of JP-4 pool fire (d = 2 m) at different relative distances y/d from the flame axis.

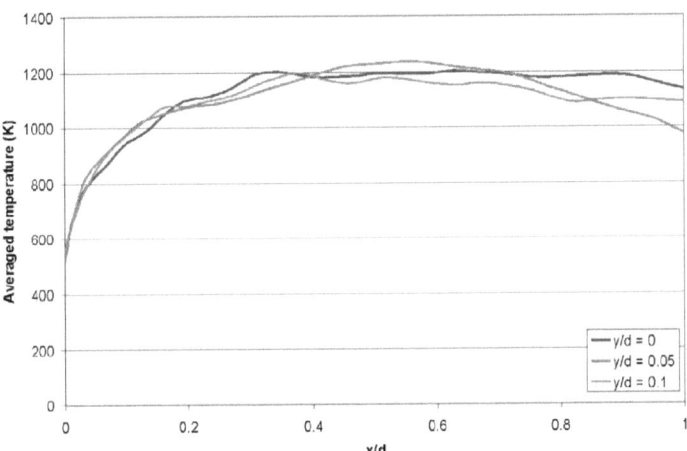

Fig. 5.4.2: Axial time averaged $\overline{T}_{CFD}$ of JP-4 pool fire (d = 8 m) at different relative distances y/d from the flame axis.

130   5. Results and discussions

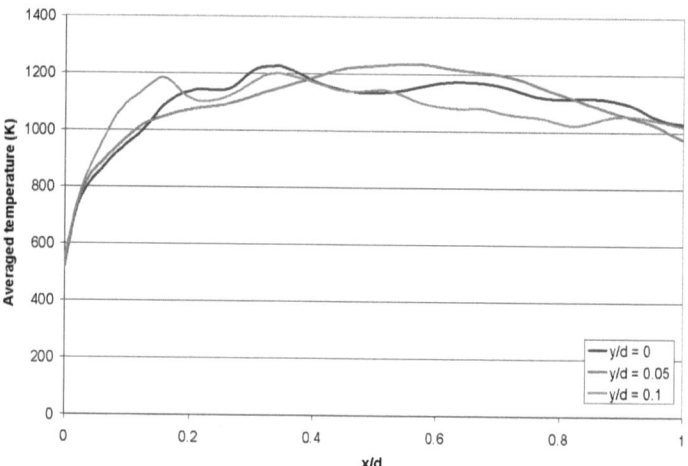

Fig. 5.4.3: Axial time averaged $\overline{T}_{CFD}$ of JP-4 pool fire (d = 16 m) at different relative distances y/d from the flame axis.

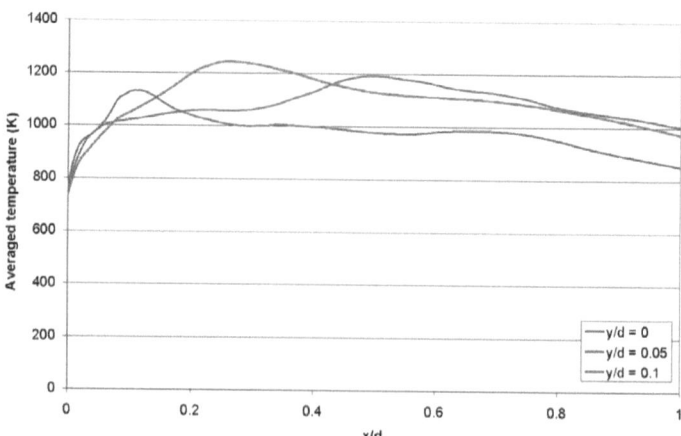

Fig. 5.4.4: Axial time averaged $\overline{T}_{CFD}$ of JP-4 pool fire (d = 25 m) at different relative distances y/d from the flame axis.

5.1 Instantaneous and time averaged flame temperatures

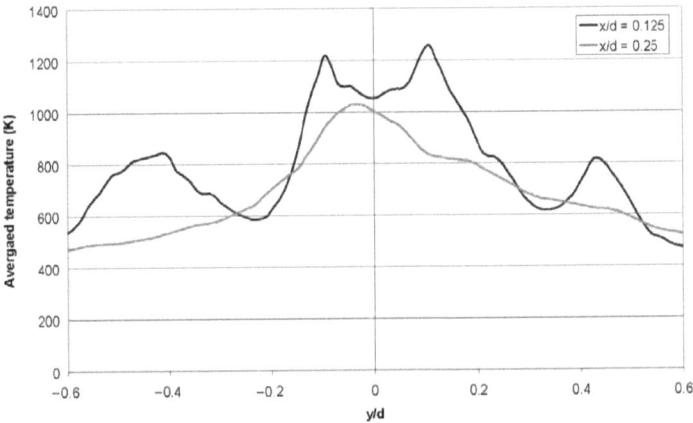

Fig. 5.4.5: CFD predicted radial profile of time averaged temperature $\overline{T}_{CFD}$ of a JP-4 pool fire (d = 25 m) at relative heights x/d = 0.125 and 0.25.

In the simulations the dependence of maximum time averaged axial temperature $\overline{T}_{max}$ (at x/d = 0.05) on d shows decreasing $\overline{T}_{max}$ with increasing d: $\overline{T}_{max}$ (d = 2 m) = 1300 K at x/d = 0.65, $\overline{T}_{max}$ (d = 8 m) = 1280 K at x/d = 0.55, $\overline{T}_{max}$ (d = 16 m) = 1250 K at x/d = 0.55 and $\overline{T}_{max}$ (d = 25 m) = 1230 K at x/d = 0.25. This is in line with the experiments of Schönbucher [3,4,7] where the average temperatures of JP-4 pool fires decrease with increasing d.

Especially, the JP-4 pool flame with 7.6 m ≤ d ≤ 15.2 m has a significantly higher measured maximum temperature of $\overline{T}_{max}$ = 1400 K whereas the simulation show a lower maximum temperature $\overline{T}_{max}$ (d = 8 m) = 1250 K (Table 5.3). Comparison of the simulated temperature profiles with the correlation in greater heights is acceptable.

In the transition zone of the flames the simulations show the same amount in decreasing the temperatures, such as the experiments [3] with exception of JP-4 pool fire with d = 2 m where this decrease is not visible for the maximum relative height of x/d = 1 as is shown on a Fig. 5.4.1. In accordance with the measured radial (horizontal) temperature profiles $\overline{T}_{exp}(r)$ for x/d, CFD predicted $\overline{T}_{CFD}(r)$ show bimodal profile at the lower part of the flame and unimodal at the upper part of the

132   5. Results and discussions

flame. For example, in a case of d = 25 m, at x/d = 0.125 bimodal $\overline{T}_{CFD}(r)$ is found, while for x/d = 0.25 already unimodal temperature profile $\overline{T}_{CFD}(r)$ exist (Fig. 5.4.5).

Table 5.3: Measured time averaged emission temperatures $\overline{T}_{em}$ of various pool fires and CFD predicted $\overline{T}_{CFD}$ of JP-4 pool fire [3,5,31].

Fuel	d (m)	$\overline{T}_{exp}$ (K)	$\overline{T}_{CFD}$ (K)	Comments
LNG	8.5 to 15	1500	-	Estimated using narrow angle radiometer data and spectral data [5]
DTBP	1.12 and 3.4	1480 and 1580	-	The maximum temperature from thermograms [31]
Gasoline	1 to 10	1240	-	
JP-4	0.1 to 10	1200	-	
JP-4	7.6 to 15.2	1400	-	The maximum temperature at the flame centerline [5]
Kerosene	1.12	1240	-	The maximum temperature from thermograms [31]
JP-4	2	1200 – 1300	1280	The maximum temperature at the flame centerline [9]
JP-4	8	-	1250	The maximum temperature at the flame centerline
JP-4	16	-	1230	The maximum temperature at the flame centerline
JP-4	20	-	1200	The maximum temperature at the flame centerline
JP-4	25	-	1200	The maximum temperature at the flame centerline

5.2 Instantaneous and time averaged Surface Emissive Power (SEP)

The SEP of a fire can be obtained by CFD simulations on *three* ways. The *first* way is based on an isosurface of constant temperature defined as a flame surface (Chapter 5.2.5). The *second* way assumes integration of many distributions of incident radiation along the z direction through the flame (Chapter 5.2.6). In the *third* way, SEP is determined by irradiance E(Δy/d, t) calculated by virtual wide angle radiometers defined and positioned at the pool rim (Chapter 5.2.7).

5.2.1 Four-step discontinuity function of temperature dependent absorption coefficient

The coupling between thermal radiation and soot reactions is described by an effective absorption coefficient $\bar{\bar{æ}}_{eff}$ of the fire. For average effective absorption coefficient $\bar{\bar{æ}}_{eff}(T)$ of JP-4 pool fires a four-step discontinuity function (Fig. 5.5) is used which includes the experimentally determined organized structures of the fire: effective reaction zones, hot spots and soot particles [3,4,7].

A detailed description of experimental procedure for determination of the parameters contained in $\bar{\bar{æ}}_{eff}(T)$ is given in [3,4,7]. The parameters are used from model OSRAMO II based on the emissive power SEP_i of the structural elements i, where i stands for the reaction zone (re), soot parcels (sp), hot spots (hs) and fuel parcels (fp). The structural elements re, sp and hs are considered as being homogeneous gas-soot particle volumes. It is assumed that emissive power originates from the reaction zones (re) of the fire, which are regarded as volume radiators.

Based on the above, modified effective absorption coefficients of the structural elements are created. Taking the differential equation for radiation transport as a basis for calculation absorption and emission, and using the Mie theory for calculating the absorption of soot particles, modified effective absorption coefficients $\bar{\bar{æ}}_{eff,i}(T)$ are derived for the structural elements i = re, sp and hs:

$$\bar{\bar{æ}}_{eff,i}(T) = \frac{36\pi f_i(n,e) X_{0.5} a_1 m_1}{c_2} \frac{c_{R,i}}{\rho_{R,i}} T_i . \tag{5.3}$$

The absorption coefficient from Eq. 5.3 can be used to define the modified, effective transmissivities $\bar{\bar{\tau}}_i(d)$:

134   5. Results and discussions

$$\bar{\tau}_i(d) = \exp(-\bar{\bar{æ}}_{\text{eff},i} d) \quad . \tag{5.4}$$

Four-step discontinuity function of the effective absorption coefficient of the fire $\bar{\bar{æ}}_{\text{eff}}(T)$ (Fig. 5.5) used in this work takes into account absorption coefficients of air and organized structures in a flame: soot parcels, hot spots and reactive zone (Chapter 2.5).

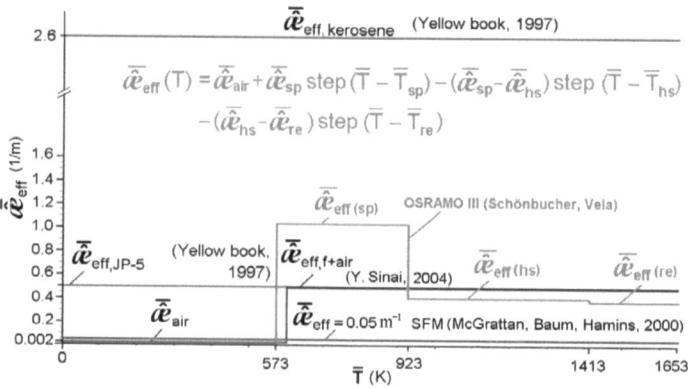

Fig. 5.5: Four-step discontinuity function of effective absorption coefficient of the pool fires.

Absorption coefficients included in $\bar{\bar{æ}}_{\text{eff}}(T)$ are: $\bar{\bar{æ}}_{\text{air}}(T)$ = 0.02 m^{-1}, $\bar{\bar{æ}}_{\text{sp}}(T)$ = 1.035 m^{-1}, $\bar{\bar{æ}}_{\text{hs}}(T)$ = 0.404 m^{-1}, $\bar{\bar{æ}}_{\text{re}}(T)$ = 0.380 m^{-1}. In Fig. 5.5 the four-step discontinuity function $\bar{\bar{æ}}_{\text{eff}}(T)$ is compared with step functions of absorption coefficients from the literature [5,14,17,25]. Sinai [17] use step function of averaged absorption coefficient $\bar{\bar{æ}}_{\text{eff}}(T)$ which contains absorption coefficient of air $\bar{\bar{æ}}_{\text{air}}(T)$ = 0.02 m^{-1} and of fuel/air reacting mixture $\bar{\bar{æ}}_{\text{f+air}}(T)$ = 0.5 m^{-1} which decrease the influence of soot absorption. Mc Grattan [14] use constant $\bar{\bar{æ}}$ = 0.05 m^{-1} of a flame (gaseous combustion mixture of fuel and air) included in SFM, neglecting a smoke blockage effect which lead to a noticeable increase of calculated surface emissive power.

5.2 Instantaneous and time averaged Surface Emissive Power (SEP)

In [25] a constant absorption coefficient $\bar{\bar{æ}} = 0.5$ m^{-1} is given for JP-5 gaseous fuel and $\bar{\bar{æ}} = 2.6$ m^{-1} for liquid kerosene fuel. The importance of use the four-step discontinuity function $\bar{\bar{æ}}_{eff}(T)$ in CFD simulation of thermal radiation from JP-4 pool fire can be seen on Fig. 5.6 where thermal radiation from the fire (in this case irradiance) is simulated by using $\bar{\bar{æ}}_{eff}(T)$ and without any absorption coefficient included. The large discrepancy in simulated results is noticeable.

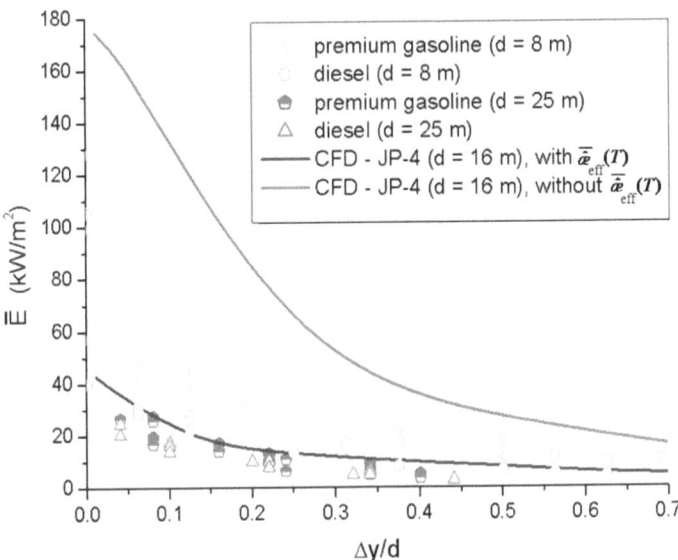

Fig. 5.6: CFD predicted time averaged irradiance $\overline{E}(\Delta y/d)$ and extrapolated $\overline{SEP}$ of JP-4 pool fire (d = 16 m) with and without absorption coefficient $\bar{\bar{æ}}_{eff}(T)$ (Fig. 5.5).

5.2.2 Thermograms

In this study the SEP is determined by evaluation of the thermograms which are measured by using a thermographic camera (Chapter 3). The experimental determination of the local distribution of surface emissive power SEP (x, y) is done by using the local temperature distribution T(x, y) from thermograms (Chapter 5.1.1).

From instantaneous temperatures $T_{i,j}$ in each pixel element i, j of the thermogram the instantaneous $SEP_{i,j}$ is determined by using the Stefan-Boltzmann-law:

$$SEP_{i,j} = \varepsilon_F \sigma T_{i,j}^4. \tag{5.5}$$

The emission level is set to $\varepsilon_F = 0.92$ [3].

The time and spatial averaged surface emissive power $\overline{SEP}$ (d), is determined by the following equations [27]:

$$\overline{SEP} \equiv <\overline{SEP}_{i,j}> = \frac{\sum_i \sum_j \overline{SEP}_{i,j} a_x}{\sum_i \sum_j a_x} \tag{5.6}$$

where

$$\overline{SEP}_{i,j} = \frac{\sum_1^{N_T} SEP_{i,j}}{N_T}. \tag{5.7}$$

where N_T is the total number of images in the series of thermograms.

Only the pixels with temperatures $T_{i,j} > 600$ K are considered for determination of SEP because the lower temperatures do not make a significant contribution to the thermal radiation of a pool fire. In Eq. (5.6) the area a_x of a pixel-matrix element is used to take into account the size and current position of a pixel in the vertical and horizontal field. Acc. to Eq. (5.6) SEP from measured thermograms are presented in Table 5.4 each for different fuels and different pool diameter [5,27,31].

The measured surface emissive power for various hydrocarbon pool fires show almost the same range of $\overline{SEP}$ depending on pool diameter d and $\overline{SEP}$ decreases with increasing d for d ≥ 1 m (Table 5.4 and 5.5, Fig. 5.11 (Chap. 5.2.5)). This effect can be explained with increasing smoke formation in these fires with increasing d and consequently decreasing of emission temperatures. Actually, in hydrocarbon pool fires (especially in higher hydrocarbons) a relatively large amount of soot (Chap. 2.5.3) is produced which leads to an increase the flame temperature and hence to an increase the thermal radiation in a fire. Due to the reason that the lack of oxygen in large hydrocarbon fires leads to the large amount of unburned cold soot, smoke which surrounds the fire, the smoke blockage effect absorb the thermal radiation from the flame inner which leads to decrease the thermal radiation to the surrounding and hence to an decrease the emission temperature at the flame surface. So, the $\overline{SEP}$ of

JP-4 pool fire show lower value than the $\overline{SEP}$ of less smoky hydrocarbon pool fires e.g. LNG, DTBP [5,27,31].

Table 5.4: Measured time averaged surface emissive power ($\overline{SEP}$) of various liquid pool fires and CFD predicted $\overline{SEP}_{CFD}$ of JP-4 pool fire [5,27,31].

Fuel	d (m)	$\overline{SEP}$ (kW/m²)	Comments
LNG	8.5 to 15	210 to 280	Estimated using narrow angle radiometer data and spectral data [5]
LNG	1 and 4	20 and 50	
DTBP	1.12 and 3.4	130 and 250	Obtained by the maximum temperature from thermograms [31]
Gasoline	1 to 10	130 to 60	Obtained by the maximum temperature from thermograms [5]
Gasoline	2.5	110	
JP-5	30	30	
Kerosene	30 to 80	10 to 25	Estimated using wide-angle radiometer data [5]
JP-4	1	100	
JP-5	1	50	
n-hexane	1	25	
n-pentane	1 and 2.5	60 and 126	

5.2.3 Histograms

The instantaneous histograms h(SEP) of SEP (Fig. 5.7a2) of a JP-4 pool fire (d = 16 m) show a local inhomogeneities and fluctuations. The time averaged histogram $\overline{h}_{CFD}$ (SEP) (Fig. 5.7a3) is obtained by averaging the instantaneous histograms h(SEP) with the number N_T.

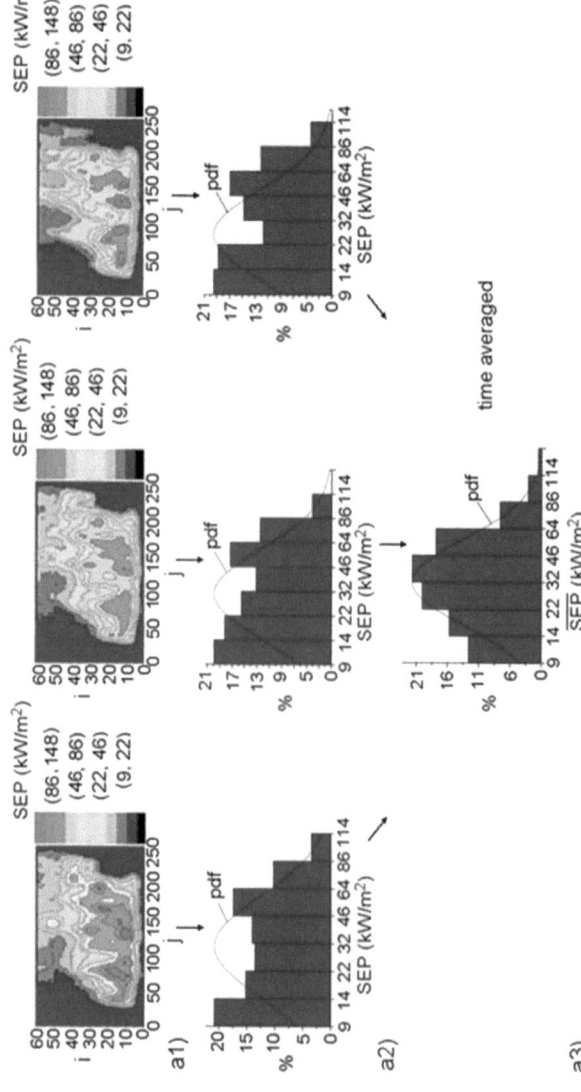

Fig. 5.7a: Evaluation of instantaneous histograms h(SEP) (a2) and time averaged histogram $\overline{h}$(SEP) of $\overline{SEP}$ (a3) of JP-4 pool fire (d = 16 m), each measured from thermograms (a1).

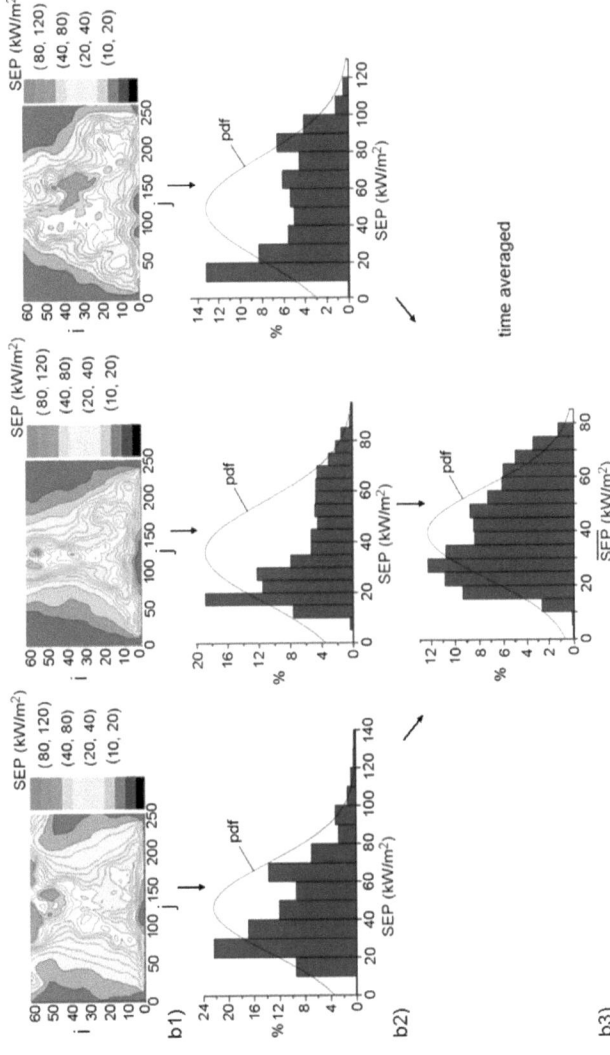

Fig. 5.7b: CFD predicted (by integration along the path length, along z-axis, in viewing direction) instantaneous histograms $h_{CFD}(SEP)$ (b2) and time averaged histogram $\overline{h}_{CFD}(SEP)$ of SEP (b3) of JP-4 pool fire (d = 16 m).

The CFD predicted instantaneous histograms h_{CFD}(SEP) (Fig. 5.7b2) obtained from thermograms (Fig. 5.7b1) are averaged over a real burning time t_b = 10 s to get time averaged histogram $\overline{h}_{CFD}$ (SEP) as shown in the Fig. 5.7b3. The maximum predicted frequency is in a range of 15 kW/m² ≤ $\overline{SEP}_{CFD}$ ≤ 50 kW/m² and maximum measured frequency is 32 kW/m² ≤ $\overline{SEP}_{exp}$ ≤ 64 kW/m² [3,4,7,27].

Low SEP values are typical for large, sooty, hydrocarbon pool fires where smoke blockage effect plays a great role in decreasing of thermal radiation of the fire to the surrounding [3,4,7,27].

5.2.4 Probability density function (pdf)

The time averaged pdf of SEP is determined based on the time averaged histogram $\overline{h}$ (SEP) (Fig. 5.7a3).

It must be noticed that the thermograms used here for visual comparison of CFD results belongs to one of the series of thermograms of JP-4 pool fire (d = 16 m) and measured $\overline{SEP}$ from thermograms vary from test to test.

The $\overline{SEP}_{exp}$ obtained from measured $\overline{SEP}$ by using statistical parameters (Table 5.5) [3] can be used for comparison with $\overline{SEP}_{CFD}$. With CFD simulation predicted $\overline{SEP}_{CFD}$ (d = 16 m) = 40 kW/m² at time averaged maximum of pdf (Fig. 5.7b3) is about 8 kW/m² greater than measured $\overline{SEP}_{exp}$ (d = 16 m) = 32 kW/m² at time averaged maximum of pdf (Fig. 5.7a3). It can be seen that the $\overline{SEP}_{CFD}$ = 40 kW/m² is about 6 kW/m² lower than the averaged $\overline{SEP}$ from series of experiments $\overline{SEP}_{exp}$ = 45.9 kW/m² (Table 5.5).

Moreover, the predicted 1st moment of pdf ($\overline{SEP}_{CFD}$) agree good with the 1st moment of thermograms obtained from log-normal pdf ($\overline{SEP}$) [3,4,7]. Due to the limited number of series of instantaneous thermograms in a visual form, only one series of N_T = 50 pictures is used for comparison with the CFD results.

Also, it should be noted that the slight wind with velocity of changeable strength and direction was present during the experiments [3,4] which may results in a more pronounced existence of hot spots and hence the heat flux received by thermograms which may have an influence on SEP of the flame and hence the discrepancy between the measured and CFD results, here in $\overline{SEP}_{exp}$ and $\overline{SEP}_{CFD}$.

Table 5.5: Averaged values and standard deviation for different d

d(m)	$(x_* = \overline{x})^1$	s	$\overline{SEP}_{exp}$ (kW/m²)	$\overline{SEP}_{CFD}$ (kW/m²)
8	− 0.043	0.190	63.8	70
16	− 0.1205	0.184	45.9	40
25	− 0.1169	0.187	38.2	38

[1] based on Gauss normal distribution [3]

5.2.5 Determination of SEP by an isosurface of flame temperature

In the first way the measured SEP is predicted by so-called incident radiation G on each grid cell placed on the given flame surface A_F.

The size G is specifically defined in CFD code as the net incident thermal radiation flux to the grid cells of the CFD simulation area [27]. More detailed, the differential radiation flux dG is calculated based on each cell of any differential surface dA in the computing differential volume dV, which is located within the computing grid for each time step. The net incident radiation G is calculated according to [27,29]:

$$G = \int_{4\pi \, sr} L(s) \, d\Omega, \tag{5.8}$$

$$G = \frac{1}{A} \int_0^{A_{F,CFD}} G \, dA \tag{5.9a}$$

$$G(A_{F,CFD}, t) \equiv SEP(A_{F,CFD}, t). \tag{5.9b}$$

The radiation intensity L in Eq. (5.8) is a result of the radiation transport equation [27,29]:

$$\frac{dL(s)}{ds} = \overline{\mathfrak{a}}_{eff}(T) L_B - \overline{\mathfrak{a}}_{eff}(T) L(s) \tag{5.10a}$$

$$L_B = \sigma T^4 / (\pi \, sr). \tag{5.10b}$$

The differential Eq. (5.10a) presents the change in L through an absorbing and emitting gray medium along a path length ds in a solid angle Ω defined around the direction of propagation **s** [27,29,83,84].

In the Eq. (5.10a) the $\bar{æ}_{eff}(T)$ is a modified absorption coefficient of the flame (Chapter 4.4.4 and 5.2.1) defined as an four step discontinuity function containing $\bar{æ}_{eff,i}(T)$ (Fig. 5.5) where i represents organized structures in a fire based on the radiation model OSRAMO II [3,4,7].

To get the SEP at the flame surface it is necessary to determine the cells lying on a isosurface which presents a realistic shape of the flame (Fig. 5.8). This can be e.g. isosurface of some constant temperature $T > T_a$.

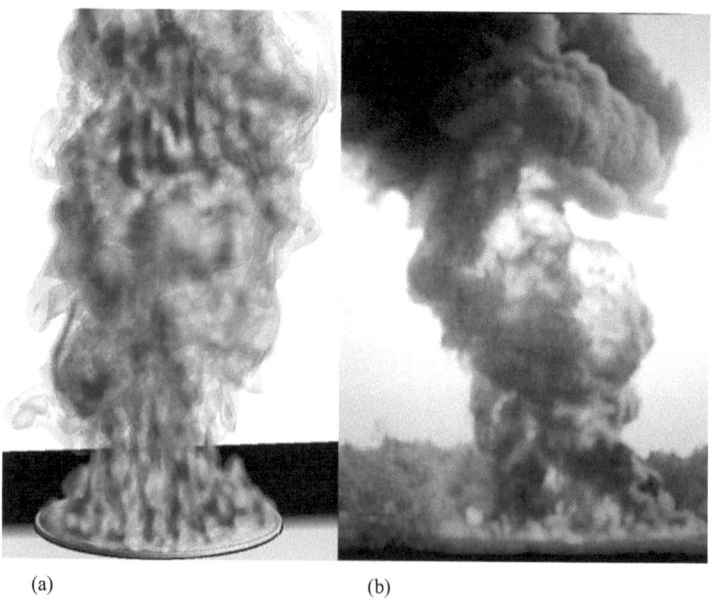

(a)   (b)

Fig. 5.8: (a) CFD predicted isosurfaces of temperatures (400 K < T < 1400 K) and (b) VIS image of a JP-4 pool fire (d = 16 m).

5.2 Instantaneous and time averaged Surface Emissive Power (SEP) 143

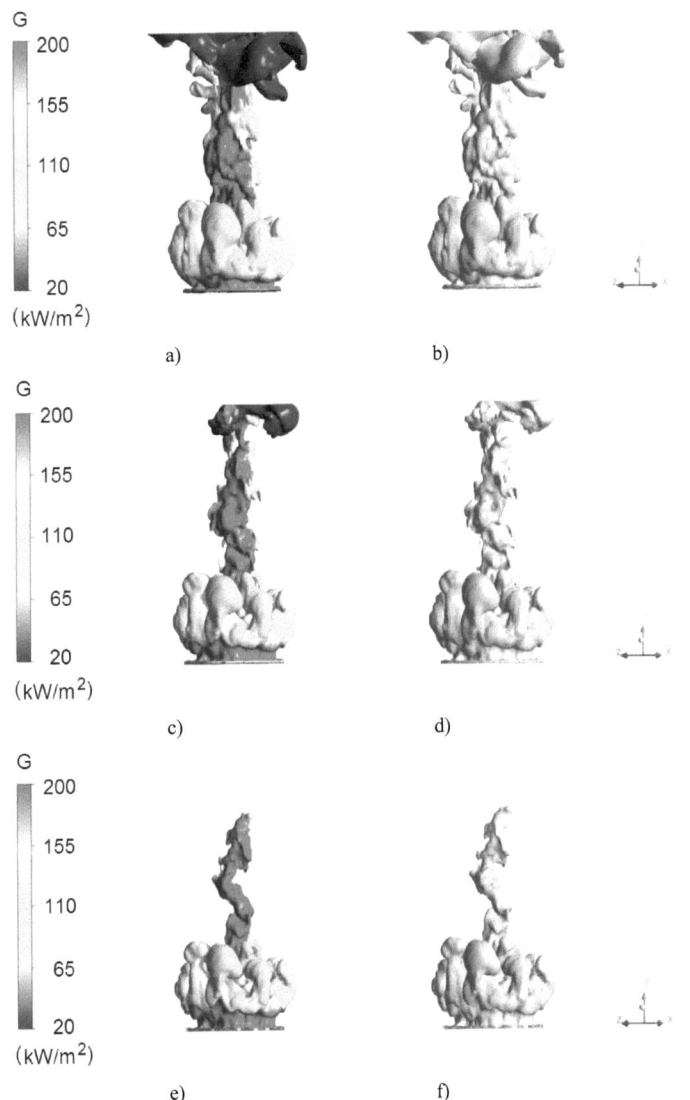

Fig. 5.9: Instantaneous isosurface of (a) T = 400 K, (c) T = 600 K, (e) T = 800 K each overlapped with G and isosurface of (b) T = 400 K, (d) T = 600 K, (f) T = 800 K of JP-4 pool fire (d = 16 m).

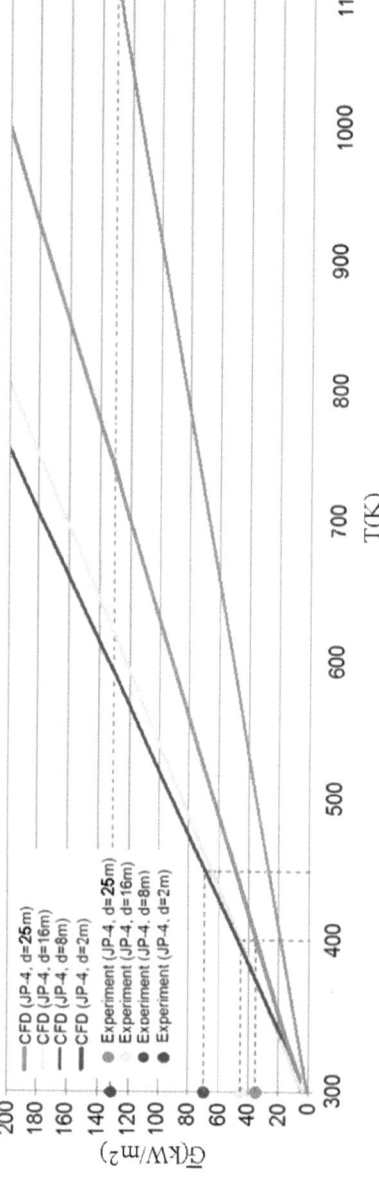

Fig. 5.10: CFD predicted $\overline{G}$ of JP-4 pool fire as a function of isosurface $A_{F,CFD}$ of constant temperature T

5.2 Instantaneous and time averaged Surface Emissive Power (SEP) 145

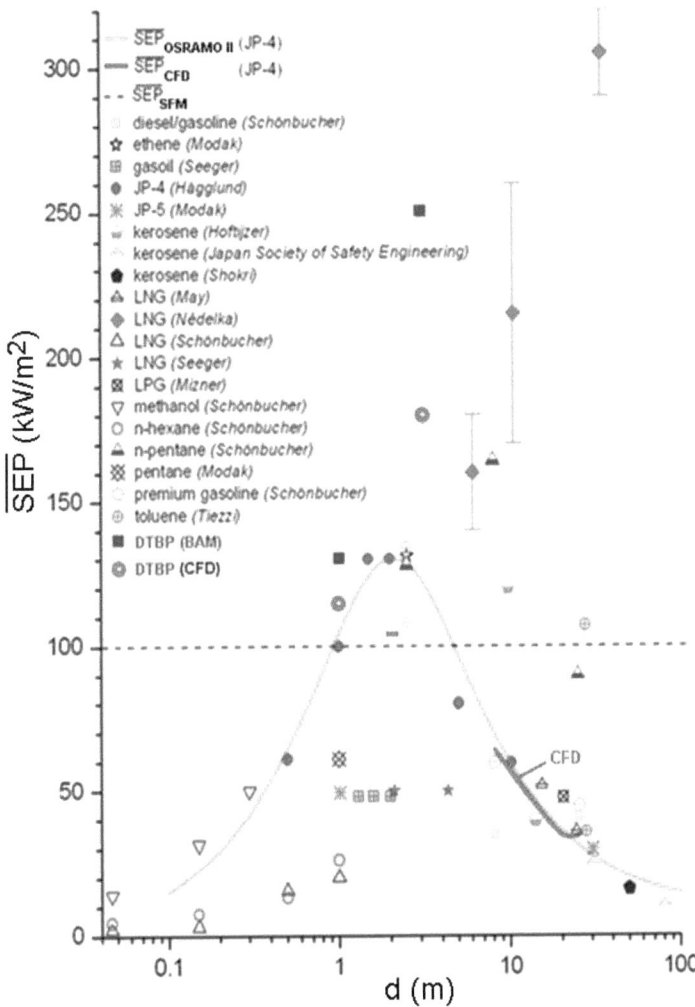

Fig. 5.11: Measured and CFD predicted $\overline{SEP}$ of hydrocarbons and DTBP pool fires as a function of d.

146   5. Results and discussions

The procedure is described in the following text:
- An instantaneous flame surface $A_{F,CFD}$ is defined as an isosurface of constant flame temperature T (e.g. Fig. 5.9b,d,f, 5.10). The isosurface temperatures T_{iso} = f(d): T_{iso} (d = 2 m) = 1100 K, T_{iso}(d = 8 m) = 450 K, T_{iso} (d = 16 m and 25 m) = 400 K are chosen to predict the time and area averaged $<\overline{SEP}>$(d) where area is represented as an isosurface of certain constant temperature (Fig. 5.9).
- The CFD calculated G(t) is averaged over the isosurface $A_{F,CFD}$ for each time interval Δt (an usually value is Δt = 0.1 s) to predict instantaneous area averaged <G(t)>.
- The <G(t)> is averaged over a burning time of t_b = 10 s which results also in time averaged $<\overline{G}> \equiv \overline{SEP}_{CFD}$ (Fig. 5. 10, 5.11). The burning time of t_b used for averaging data is defined as a time starts at t = 10 s when the flame is developed and ends at 20 s limited by the CPU time. It is assumed that the flame during that burning time of 10 s show a real burning.

5.2.6 Integration of incident radiation G for determination of SEP

In the second way SEP is predicted based on the integration of many G distributions in x,y-direction along the z direction through the flame (Fig. 5.12).

With the CFD simulation predicted SEP_{CFD} (x, y, t) distribution is defined through integration of instantaneous G (x, y, t) distributions within many (e.g. 50) vertical parallel planes to the direction (z-axis) on the path length s:

$$SEP(x,y,t) = \int_0^s G(x,y,t)\,dz. \qquad (5.11)$$

The path length s is dependent on the dynamic properties of fire and in this work used range of s is $d/4 \leq s \leq d/2$. Integration in Eq. (5.11) is done for each time step Δt = 0.1 s, giving the instantaneous histograms of SEP and calculated probability density function (pdf) of SEP (Fig. 5.7b2, Chapter 5.2.4).

5.2 Instantaneous and time averaged Surface Emissive Power (SEP) 147

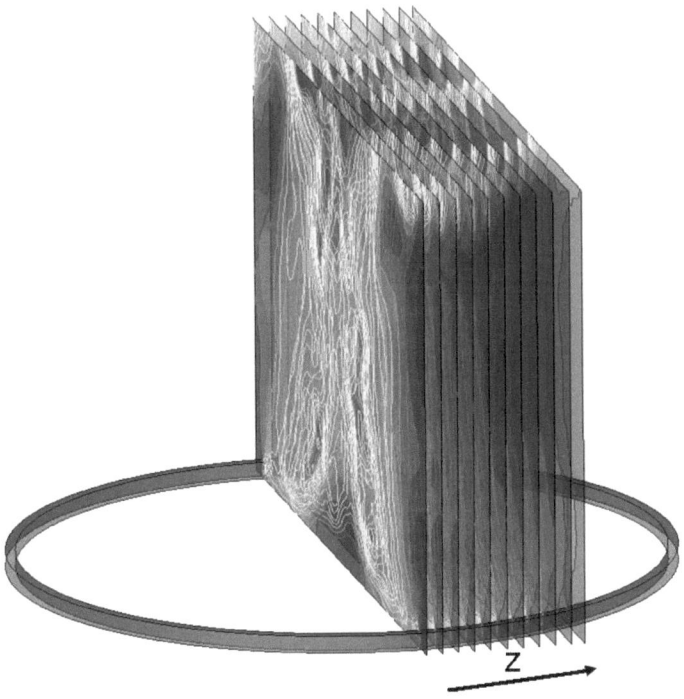

Fig. 5.12: Instantaneous distribution of G(x,y,t) as e.g. 11 parallel planes perpendicular to the line of sight (z-axis).

5.2.7 Determination of SEP by irradiance as a function of distance

In the third way, $\overline{SEP}_{CFD}(d)$ of a fire is predicted by the calculated irradiance $\overline{E}_{CFD}(\Delta y/d, d)$ with virtual wide angle radiometers defined and positioned at the pool rim (Eq. (5.12), Chap 5.3.2) at the relative distance $\Delta y/d = 0$, as shown on Fig. 5.13.

By the wide angle virtual radiometers the irradiance $E_{CFD}(\Delta y/d, t)$ depending on the relative distance $\Delta y/d$ (Fig. 5.13, Chapter 5.3.2) and time t is predicted. The $E_{CFD}(\Delta y/d, t)$ are averaged over a burning time t_b to get $\overline{E}_{CFD}(\Delta y/d, d)$:

$$\overline{SEP}_{CFD} \equiv \overline{E}_{CFD}(\Delta y/d=0, d). \tag{5.12}$$

The agreement between prediction and radiometer measurements at relative radial distance $\Delta y/d = 0$ for JP-4 pool fire ($d = 2$ m, 8 m, 16 m and 25 m) (Table 5.6 and Fig. 5.14) is very good.

Table 5.6: Measured [3,5] and CFD predicted $\overline{SEP}$ of different fuels with different d

Fuel	d (m)	$\overline{SEP}_{exp}$ (kW/m²)	$\overline{SEP}_{CFD}$ (kW/m²)
LNG	8.5 to 15	210 to 280	
LNG	1 and 4	20 and 50	
Gasoline	1 to 10	130 to 60 (max)	
Gasoline	2.5	110	
JP-5	30	30	
Kerosene	30 to 80	10 to 25	
JP-4	1	100	
JP-4	2	130	105
JP-4	8	70	70
JP-4	16 m	45	45
JP-4	20 m	31	32
JP-4	25 m	35	35

5.3 Instantaneous and time averaged irradiance

5.3.1. Virtual radiometers

In the CFD study virtual radiometers or receiving elements are defined in points placed at different relative distances $\Delta y/d$ from the pool rim as in the experiments [3]. Two radiometers were defined at the each point at different heights: h = 0.5 m and 1 m. For each virtual radiometer a view factor is defined. The radiometers have a temperature of T = 0 K, absorption and a sensitivity of 100% assigned, i.e. no heat loss to the recipients is considered. The opening angle is 180°. In the resulting field of view of the receiving element, from the radiometer coordinates, 8 · 8 rays are sent back, along the path on which the radiation is integrated [27]. In CFD code the angular calibration table that is used is printed out, for each radiometer location, the following is written: location, direction, temperature, flux [27,28].

5.3.2. Prediction of irradiance

In safety, the irradiance plays an important role. The influence of the irradiance to an object exposed to the fire depends on the intensity and duration of spontaneous combustion. In humans, the received radiation depends on the intensity at a shorter duration of exposure to pain, injury or even death. More details about critical irradiation on human skin can be found in [7].

In CFD simulation a net radiation flux from the flame is received in a certain computational cell where the virtual radiometer is defined. The calculation of irradiance E ($\Delta y/d$, d) is done by using defined directions along light path s. In the resulting field of view of virtual receiving element of the respective radiometer the radiation flux q_r is integrated along the coordinates of $8 \cdot 8$ beams [27,28]:

$$q_r(\Delta y / d) = \int_{4\pi\Omega_0} L(\Delta y / d, s) s \, d\Omega \equiv E_{CFD}(\Delta y / d). \tag{5.13}$$

In the simulation of JP-4 pool fires (d = 2 m, 8 m, 16 m and 25 m) irradiances E ($\Delta y/d$, d, t) are calculated at different relative distances to the pool edge and at a height of x = 0.5 m and 1 m, depending on time.

Irradiance received by an object strongly depends on the distance from the fire and exposure time. At a constant distance occur over time at very high fluctuation of E ($\Delta y/d$, t) (Fig. 5.13).

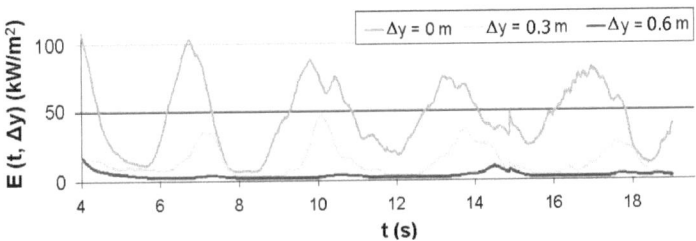

Fig. 5.13: Time dependent irradiance E ($\Delta y/d$) of JP-4 pool fire (d = 16 m) at different distances Δy from the pool rim.

Fig. 5.13 show an example of time histories of calculated instantaneous irradiance E ($\Delta y/d$, t) for JP-4 pool fires (d = 16 m) in different radial distances $\Delta y/d$ from the pool rim at the height of x = 1 m. All calculated curves reflect the dynamics of the flames. Because of the pulsing behavior the surface emissive power SEP(t) is subjected to periodic changes in intensity which also result in the virtual receiver elements with clearly divergent maxima and minima of the irradiance E(t). These temporal changes of E(t) are supported by experimental results [3].

Irradiances $\overline{E}(\Delta y/d, d)$ are measured at different relative distances $\Delta y/d$ from the pool rim for JP-4, n-pentane and regular gasoline pool fires (d = 2 m, 8 m, 16 m and 25 m) [3]. The measured averaged irradiances $\overline{E}_{exp}(\Delta y/d)$ of different fires decrease with $\Delta y/d$ (Fig. 5.14).

To predict the time averaged irradiance $\overline{E}_{CFD}(\Delta y/d)$, the time dependent $E_{CFD}(\Delta y/d, t)$ from Eq. (5.13) is averaged over the burning time t_b = 10 s. With increasing relative distance $\Delta y/d$ the time averaged irradiance of e.g. of JP-4 pool fire (d = 16 m) decreases from $\overline{E}_{CFD}(\Delta y/d = 0) = 46$ kW/m^2 to $\overline{E}_{CFD}(\Delta y/d = 0.2) = 15$ kW/m^2 and $\overline{E}_{CFD}(\Delta y/d = 0.4) = 11$ kW/m^2 (Fig. 5.14).

The predicted $\overline{E}_{CFD}(\Delta y/d)$ profiles (Fig. 5.14) are in good agreement with the measured $\overline{E}_{exp}(\Delta y/d)$ values for the JP-4 pool fire with d = 8 m, 16 m, 25 m [3] whereas $\overline{E}_{CFD}(\Delta y/d)$ of JP-4 pool fire with d = 2 m relatively agrees with $\overline{E}_{exp}(\Delta y/d)$ for $\Delta y/d = 0$ but over predicts $\overline{E}_{exp}(\Delta y/d)$ for a larger distance. Because of the less number of experimental data [3] this discrepancy is questionable.

Due to the limited extension of the computational mesh $\overline{E}_{CFD}(\Delta y/d)$ are not calculated for larger distances $\Delta y/d > 1.1$.

Both in the simulations as well as in the experiment a significant decrease in irradiance with increasing distance from the pool rim is determined. The significantly higher measured $\overline{SEP}$ of the smaller JP-4 pool flame $\overline{SEP}(d = 2\text{ m}) = 130$ kW/m^2 and CFD predicted $\overline{SEP}_{CFD}(d = 2\text{ m}) = 105$ kW/m^2, means however, a higher irradiance than in a case of larger JP-4 pool flame. The irradiance dependence on pool diameter is reflected in the measurements [3].

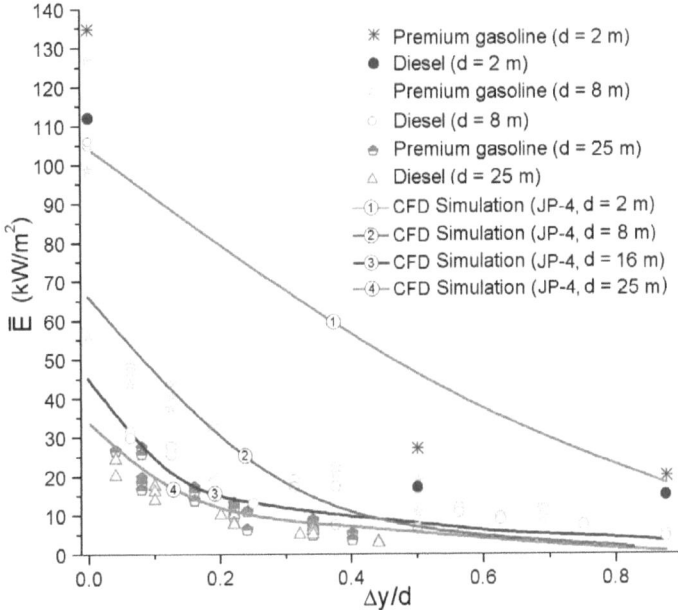

Fig. 5.14: Measured $\overline{E}_{exp}(\Delta y/d)$ and CFD predicted time averaged irradiances $\overline{E}_{CFD}(\Delta y/d)$ from the large JP-4 pool fires as a function of relative distance $\Delta y/d$ from the pool rim.

The decrease in the calculated irradiance with increasing distance from the pool fire predicted in the simulations $\overline{E}_{CFD}(\Delta y/d)$ and the pool diameter dependence are in agreement with experiments in most cases (Fig. 5.14). That shows an ability of CFD simulation in determination of an appropriate safety distance for the protected objects. For calculation of thermal hazard a minimum value of $\overline{E}(\Delta y/d) = 10$ kW/m² is used. The maximum irradiance $\overline{E}_{max}(\Delta y/d)$ is reached in a very short time period, so taking into account a time exposure, a time averaged irradiance $\overline{E}(\Delta y/d)$ is more interesting from the safety point of view. Generally, the excess from the averaged $\overline{E}(\Delta y/d)$, the rapid increase to a temporary maximum and the subsequent reduction to a temporary minimum takes a time period of 2 s ≤ Δt ≤ 4 s. This duration is sufficient to injuries to persons, as the critical exposure duration is 3 s. For technically protection of important objects, these short-term maxima, is not significant because the critical exposure period is usually much longer [7].

5.4 Wind influence

5.4.1 Flame height, flame tilt, flame drag

Flame height, flame tilt

The flame height is determined by means of the visible images obtained from the VHS video recordings. The camera used to determine the tilt and length of a flame was placed perpendicular to the predominant wind direction: those tests in which the wind velocity was nearly constant and parallel of camera view were selected. For each test, a portion of film corresponding to the stationary state was selected, digitalized, and divided into a sequence of digital images at 25 frames per second. An algorithm was developed to allow the maximum height of the visible luminous flame to be selected for each frame in the sequence [3,4,27].

In CFD simulation a flame tilt is determined by axial profile of temperature using a maximum temperature, e.g. T_{max} of pulsation flame zone to determine the flame height. The distance of maximum point of the flame height from the flame axis is used to determine the tilt angle from the vertical.

Results obtained from the simulations with an additional wind flow shows that the flame is tilted to the side, but the frequency of the formation of vortices, and thus the fluctuation of the irradiance, remains nearly unchanged unless the flame is tilted more than 30 degrees.

CFD results (Fig. 5.15a,b - 5.17a,b) show that the influence of wind leads to:

- Flame tilt from the vertical
- Downstream movement and stretch of vortices near the pool rim parallel to the ground
- Formation of large counter rotating vortices on the downwind side of the flame
- Increasing of instantaneous temperatures and irradiances at the downwind side of the fire.

In CFD simulation with JP-4 pool fire with d = 20 m with different cross wind velocities the following results are obtained:

- In a case of wind velocity of $u_w = 0.7$ m/s no discernible flame tilt is observed neither in experiments neither in CFD results.
- With increasing wind velocity to $u_w = 2.3$ m/s a flame tilt from the vertical became significant in CFD results and comparable with experiments, after 2 s of simulated time. The tilt is noticeable at first near above the pool rim, in stretching the vortices and forming a flame drag. The CFD simulation show two counter rotating vortices at the leeward side of the fire as observed in experiment. The determination of the flame tilt, drag and vortices is done by prediction of position of the maximum temperatures and the flow field in the fire.
- With increasing wind velocity to $u_w = 5.5$ m/s to 10 m/s a flame tilt became significant in CFD immediately after 1 s of simulated time. The strong tilt is noticeable also in a fire plume at $u_w = 5.5$ m/s and the flame drag show spreading of the fire across pool for 2d in comparison with a calm condition.

CFD results shows (Fig. 5.15a,b) isotherms of temperature and isosurfaces of constant temperatures (400 K ≤ T ≤ 1000 K) predicted with RANS. Fig. 5.15a,b presents CFD predicted transient isotherms in the xy-plane and isosurfaces of constant temperatures of a JP-4 pool fire (d = 20 m) under the cross wind with velocity of u_w = 4.5 m/s in two different times. The figures show influence of the cross wind on the flame tilt and temperature distribution of the fire. Relative to the ambient air rapidly rising flame gases by transmission from the surrounding air change the vortex structures. The vortices are created directly on the edge of the pool and rise up with the times. At the time t_1 = 6 s is, for example, a vortex structure such that the center is located with the coordinates $x_1(t_1)$ = 12 m above the right edge of the pool and $y_1(t_1)$ = 16 m from the flames axis (Fig. 5.15a). The origin lies, however in all the simulations in the middle of the pool. At t = 10 s the eddy is moved to the much higher position at $x_2(t_2)$ = 32 m and $y_2(t_2)$ = 23 m (Fig. 5.15b).

An increase in the vortex diameter is recognizable at the earlier stage at $t_1 < 6$ s. The movement and resizing of vortices due to the wind influence reflect the isotherms of the flame. By the rising of eddy thermal energy is increasingly perpendicular to flames axis and at the time t_2 the areas of higher temperatures occur in larger intervals inside the vortices.

154   5 Results and discussions

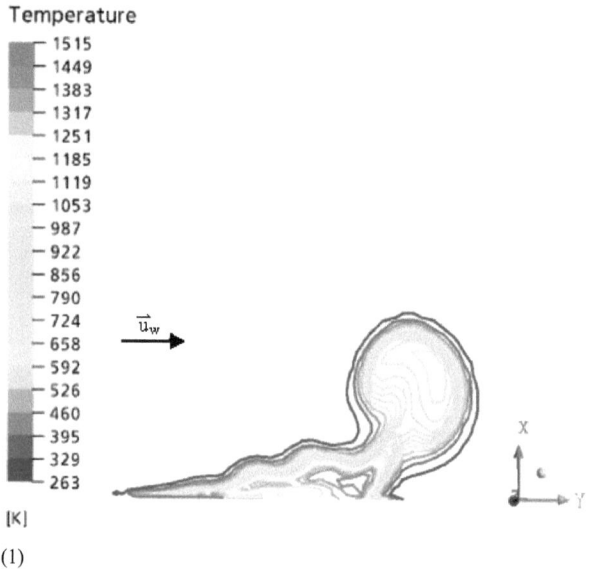

(1)

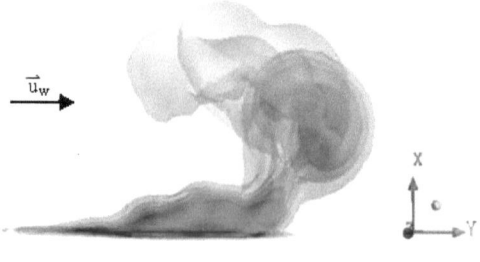

(2)

Fig. 5.15a: CFD predicted: isotherms (1) and isosurfaces (2) of constant flame temperatures (400 K ≤ T ≤ 1000 K) of a JP-4 pool fire (d = 20 m) under the cross-wind of u_w = 4.5 m/s at t = 6 s.

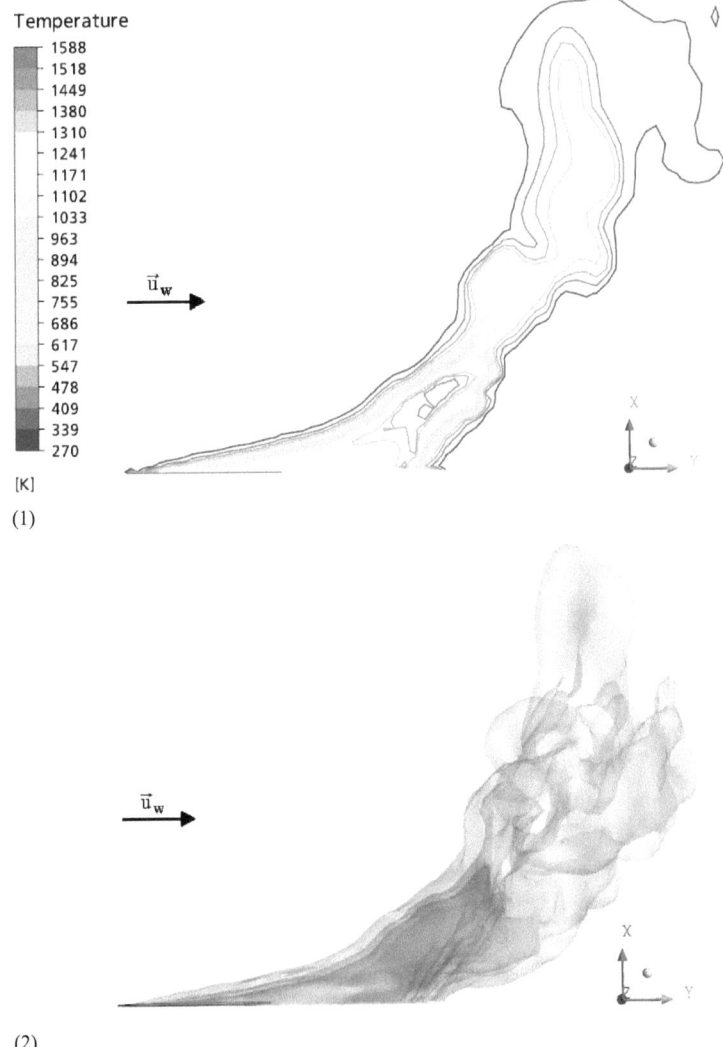

Fig. 5.15b: CFD predicted: isotherms (1) and isosurfaces (2) of constant flame temperatures (400 K ≤ T ≤ 1000 K) of a JP-4 pool fire (d = 20 m) under the cross-wind of u_w = 4.5 m/s at t = 10 s.

By vortex propagation oxygen (air) is transported into the flame, so that the flame is directed to areas where the vortex combustion is favorable (Chapter 2.10), so more thermal energy is released. This makes the very high temperatures of $T \geq 1300$ K noticeable. The periodic rise of vortices influences the flame pulsation [28]. The maximum speeds in axial direction occur at all flames near the flames axis.

The crosswind produces an effect along the downwind side of the pool in the form of counter rotating-vortices (Fig. 5.15a,b-5.17a,b). The counter-rotating vortices that appeared in the fire with 1.4 m/s wind speed are also observed with the 7.2 m/s and 10 m/s wind speed (Fig. 5.17a,b). They follow the stream-wise direction and are close to the ground. The use of RANS model in CFD simulation successfully predicts this effect. These structures have a time mean definition so RANS formulation could predict their existence.

Fig. 5.17b shows vertical cross-sections of isotherms of temperature through the calculation domain near the leeward edge of the pool predicted with RANS. The RANS calculations are qualitatively in a good agreement with the test photographs [3,4,9,10,16,19].

The CFD results show that the wind speed of 5 – 7 m/s tilt the flame much farther and a portion of the flame zone lays on the ground beyond the leeward edge of the pool as it is observed in the experiments [9,11]. CFD results show that the flame volume attaches to the ground for more than an additional pool diameter downwind of the pool as it is observed in test photographs [9]. The flame footprint, or the area near the ground, can also be visualized by looking at the temperatures in a horizontal plane just above the ground (Fig. 5.17a). A horizontal plane through the CFD predicted results (Fig. 5.17a) indicates the two columns as the strong counter-rotating vortices. Both simulations and the photographs [9,16] indicate an interior region between the columnar vortices free of flames (Fig 5.17a,b).

Even with a k-ε model, the initial puffing of fire that form during the initial transient simulation, as the fire plume establishes itself after ignition, is always calculated. The rollup of this initial vortex is not suppressed by eddy viscosity in the k-ε model.

Fig. 5.16: (a) Thermogram, (b) VIS image and (c) CFD predicted isosurfaces of temperature (400 K < T < 1400 K) of a JP-4 pool fire (d = 25 m) under the influence of the cross-wind (u_w = 4.5 m/s).

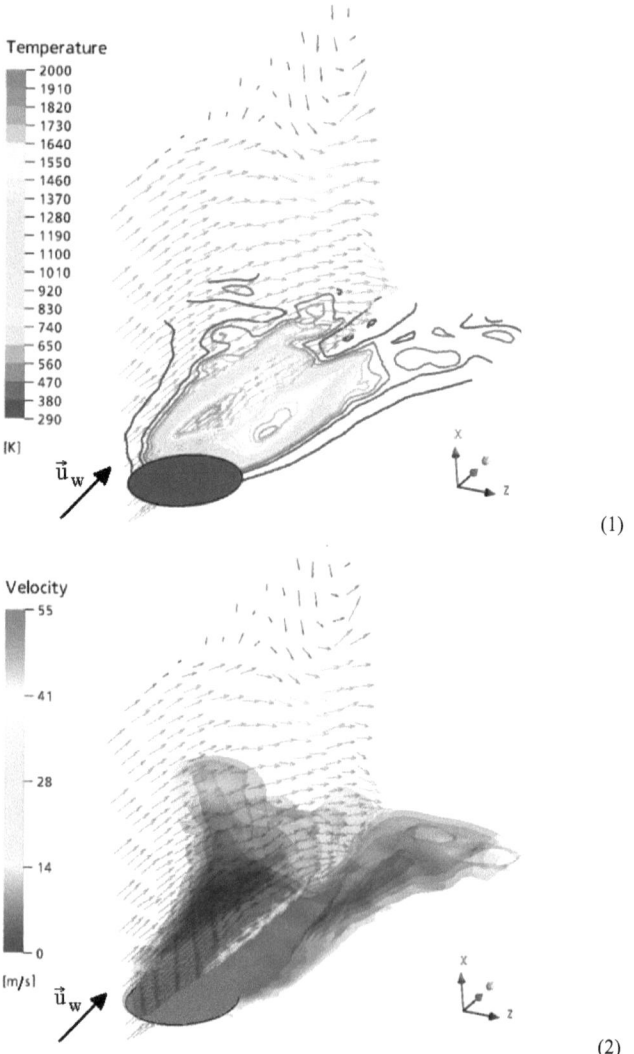

Fig. 5.17a: CFD predicted counter-rotating vortices by (1) isotherms and (2) isosurfaces of temperatures (400 K < T < 1400 K) of JP-4 pool fire (d = 20 m) on the horizontal plane at the ground level under the influence of the cross-wind (u_w = 7.2 m/s) at t = 18 s.

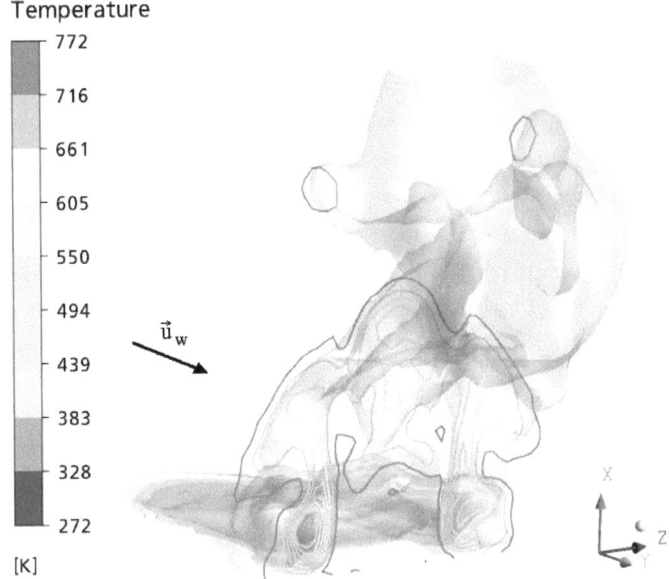

Fig. 5.17b: CFD predicted counter-rotating vortices by isotherms (on the cross-plane) and isosurfaces of temperatures (400 K < T < 1400 K) of JP-4 pool fire (d = 20 m) under the influence of the cross-wind (u_w = 10 m/s) at t = 11 s.

CFD simulation of the wind influence on unconfined large kerosene pool fire (d = 20 m) is done by Sinai [19] who used CFDS-FLOW3D code. The CFD data have been compared with experimental data of Shell Research Ltd at British Gas Test Site Spadeadam, Cumbria. He investigated a role of a pool shape and ambient turbulence on the behavior of the fire under the cross-wind (in introduction). His CFD results show flame tilt ranged from 43° to 56° which does not bracket the measured value of 40° and predictions of flame base length ranged from 20 m to 40 m, compared with the observed value of 20 m to 50 m [9,19].

In this work CFD predicted flame tilt of JP-4 pool fire (d = 2 m and 20 m) show agreement with experimental data from [9,11,19].

The prediction of the flame tilt for d = 20 m pool fire by using profiles of maximum temperatures show increasing of a flame tilt from a vertical with increasing wind

160   5 Results and discussions

velocity u_w, from 20° for $u_w = 1.4$ m/s to 70° for $u_w = 10$ m/s whereas the tilt of 80° is predicted when u_w reaches 16 m/s.

The CFD predicted flame tilt and a flame drag by means of averaged data for different wind velocities compared with calculated flame tilt and flame drag are presented in Table 5.9. The results are obtained for each time step Δt (e.g. $\Delta t = 0.1$ s) and averaged over the burning time of $t_b = 10$ s.

The CFD predicted results agree well with the experimentally obtained photographs [3,4,9,10,16,19] and calculated data (Table 5.9). The empirical correlations used for comparison with CFD predicted flame tilt and a flame drag (Table 5.7 and 5.8) are based on the experiments with gasoline and diesel pool fires [20] and small scale fires [25].

Flame tilt

$$\cos\Theta = \begin{cases} 1 & , \text{for } \bar{u}_w^* < 1 \\ a_1 (\bar{u}_w^*)^{b_1} & , \text{for } \bar{u}_w^* \geq 1 \end{cases} \quad (5.14a)$$

with $\bar{u}_w^* = \bar{u}_w / \bar{u}_c$ (5.14b)

and $\bar{u}_c = (g\overline{\dot{m}_f''}d / \rho_v)^{1/3} \approx (g\overline{\dot{m}_f''}d / \rho_a)^{1/3}$ (5.14c)

where $\bar{u}_w^*$ is a wind velocity measured at height of 1.6 m;

Table 5.7: Empirical correlations [7] for calculation the flame tilt used in Eq. 5.14a.

Correlation	a_1	b_1	Comments
AGA	1	−0.5	measured on LNG pool fires [5]
Thomas	0.7	−0.49	measured on wood fires [42]
Moorhouse	0.86	−0.25	measured on large cylindrical LNG pool fires, $\bar{u}_w^* = \bar{u}_w^*(10)$ [44]
Muñoz	0.96	−0.26	measured on gasoline and diesel pool fires [20]

Flame drag

$$\bar{d}_w / d = c^1 (Fr)^{d^1} (\rho_v / \rho_a)^{e^1}. \quad (5.15)$$

5.4 Wind influence

Table 5.8: Empirical correlations [25] for calculation the flame drag used in Eq. 5.15.

Correlation	c_1	d_1	e_1	Comments
-	1.6	0.061	0	conical flame
-	1.5	0.069	0	cylindrical flame
Sliepcevich	2.1	0.21	0.48	

The correlation of Munoz (Table 5.7) is used for comparison with CFD predicted flame tilt and the flame drag calculated with correlation of Sliepcevich (Table 5.8) is compared with CFD predicted flame drag (Table 5.9). The CFD predicted flame tilts (Table 5.9) agree with experimental and calculated data, except for d = 20 m JP-4 pool fire under the small wind velocity $u_w \leq 1.4$ m/s where calculation neglect the flame tilt.

Table 5.9: Measured, calculated and CFD predicted flame tilt and a flame drag

d (m)	u_w (m/s)	Tilt (exp)	Tilt (calc)	Tilt (CFD)	Drag (exp)	Drag (calc)	Drag (CFD)
2	4.5	60°	63°	60°	1.8	1.6	2.5
2	10	70°	72°	80°	2.0	1.6	2.6
2	16	70°	76°	80°	2.0	1.6	2.8
20	0.7	20°	0°	20°	1.1	1.0	1.1
20	1.4	20°	0°	20°	1.1	1.0	1.1
20	2.3	30°	20°	30°	1.2	2.0	1.4
20	4.5	40°	48°	50°	1.5	2.0	1.5
20	5.5	50°	52°	60°	2.0	2.0	1.6
20	7.2	70°	58°	60°	2.0	2.0	1.8
20	10	n.a.	63°	70°	n.a.	2.0	2.0
20	16	n.a.	69°	80°	n.a.	2.0	2.5

CFD over predicts flame drag in a case of d = 2 m JP-4 pool fire whereas for d = 20 m agree with the experimental data. Calculated flame drag show constant value which refers to the averaged value based on the experiments on small pool fires or large

LNG fires [25] which can explain the discrepancy between calculated and CFD predicted results.

5.4.2 Wind influence on surface emissive power (SEP), irradiance (E), temperature, flow velocity

Wind influence on surface emissive power (SEP) and irradiance (E)

To illustrate the relationship between air flow and temperature field, SEP and irradiance, the CFD simulations of JP-4 pool fires (d = 2 m and 20 m) under the influence of cross wind with different wind velocities (u_w = 1.4 m/s, 2.3 m/s, 4 m/s, 4.5 m/s, 10 m/s and 16 m/s) are performed.

The following flow processes can be found both in the simulated JP-4 pool fires comparable with experiments on JP-4 and JP-8 pool fire [9,11,19].

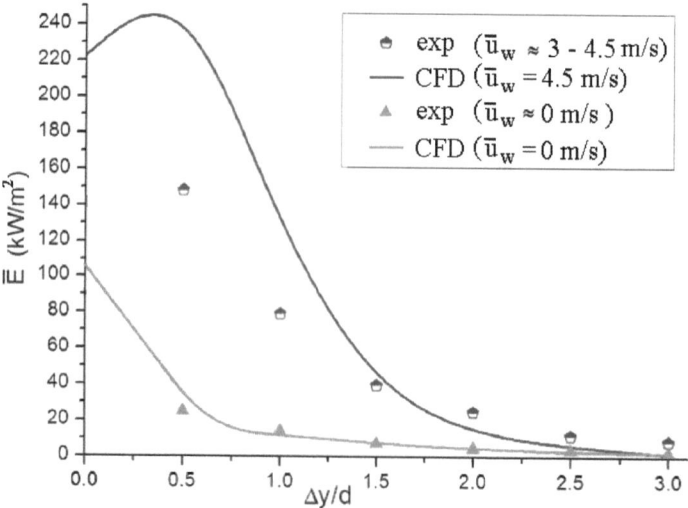

Fig. 5.18: Measured [11] and CFD predicted irradiances $\overline{E}(\Delta y/d)$ at different relative distances $\Delta y/d$ from the pool rim of a JP-4 pool fire (d = 2 m) under the influence of cross wind with (u_w = 4.5 m/s).

It is found that the wind causes significant radiation increase in the downwind direction of a fire (Fig. 5.18, 5.19) and affects the radiation level slightly in the upwind and cross-wind directions as it is observed in experiments [9,11]. CFD predicted SEP and E downwind of the pool increases with wind speed and reaches a maximum at the wind speed of about $u_w = 10$ m/s which is close to the experimentally [11] obtained value of 6.7 m/s $\leq u_w \leq 8.9$ m/s and decreases at higher speeds. The large values of SEP and E downwind of the fire is influenced also with the flame tilt.

Due to the smoke blockage, intense heating and improper viewing the radiation measurements near the fire is difficult so the experimental results show radiation measured at $\Delta y/d \geq 0.5$ [11] (Fig. 5.18). This can explain the discrepancy between predicted and measured $\overline{E}(\Delta y/d)$ at the closer distance to the pool rim (e.g. for $\Delta y/d \leq 0.5$ on the Fig. 5.18).

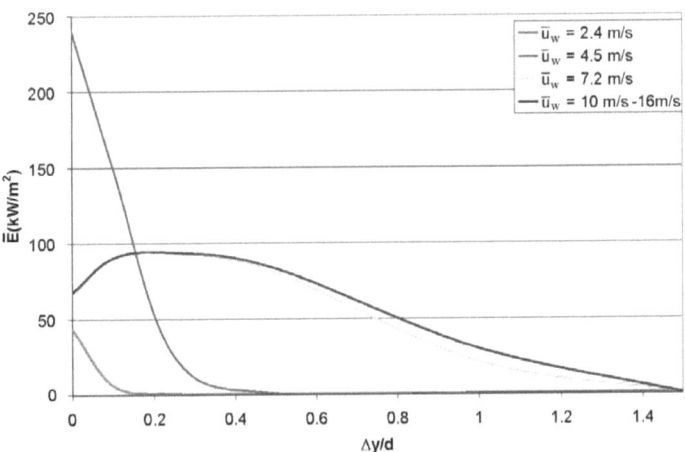

Fig. 5.19: CFD predicted time-averaged irradiances $\overline{E}(\Delta y/d)$ at different relative distances $\Delta y/d$ from the pool rim of a JP-4 pool fire (d = 20 m) under the influence of the cross-wind with different wind velocities.

Wind influence on temperature

Increasing the wind velocity up to e.g. u_w = 10 m/s, due to the flame tilt and drag the predicted maximum time averaged axial temperatures $\overline{T}_{max,CFD}$ of JP-4 pool fires are found at larger relative distances y/d > 1 from the flame centerline as shown on Fig. 5.20-5.23.

In a case of JP-4 pool fire (d = 20 m), with increasing the wind velocity up to e.g. u_w = 10 m/s, the predicted maximum time averaged axial temperature $\overline{T}_{max,CFD}$ is found at the axial distance x/d = 0.08 and radial distance $\Delta y/d$ = 1 (y/d = 1.5) from the pool rim (Fig. 5.23).

CFD simulation shows that under the influence of the wind with u_w = 10 m/s the fire diameter (d = 20 m) extends for an additional d (Table 5.9, Fig. 5.22 and 5.23). On the Fig. 5.23 is shown that the maximum time averaged axial temperature $\overline{T}_{max,CFD}$ is found at the relative distance of y/d = 1.5 from the flame centerline which means that the flame tilt from the vertical is about 70° and a flame drag is 2 (Tab. 5.9). At the flame centerline the very high temperatures can still be found at a very low axial distance from the pool (x/d = 0.025) (Fig. 5.23).

In the Fig. 5.20-5.21 CFD predicted isotherms and isosurfaces of temperatures (400 K < T < 1400 K) show instantaneous flame tilt and drag of JP-4 pool fire with d = 2 m and 25 m under the wind velocity of u_w = 4.5 m/s, compared with VIS images.

The maximum time averaged axial temperature $\overline{T}_{max,CFD}$ in a case of lower wind condition (e.g. u_w = 4.5 m/s) is found at the relative distance of y/d = 2 and 1 (for d = 2 m and 20 m) from the flame centerline which means that the flame tilts from the vertical are about 60° and 50° and the flame drags are 2.5 and 1.5 (for d = 2 m and 20 m) (Tab. 5.9).

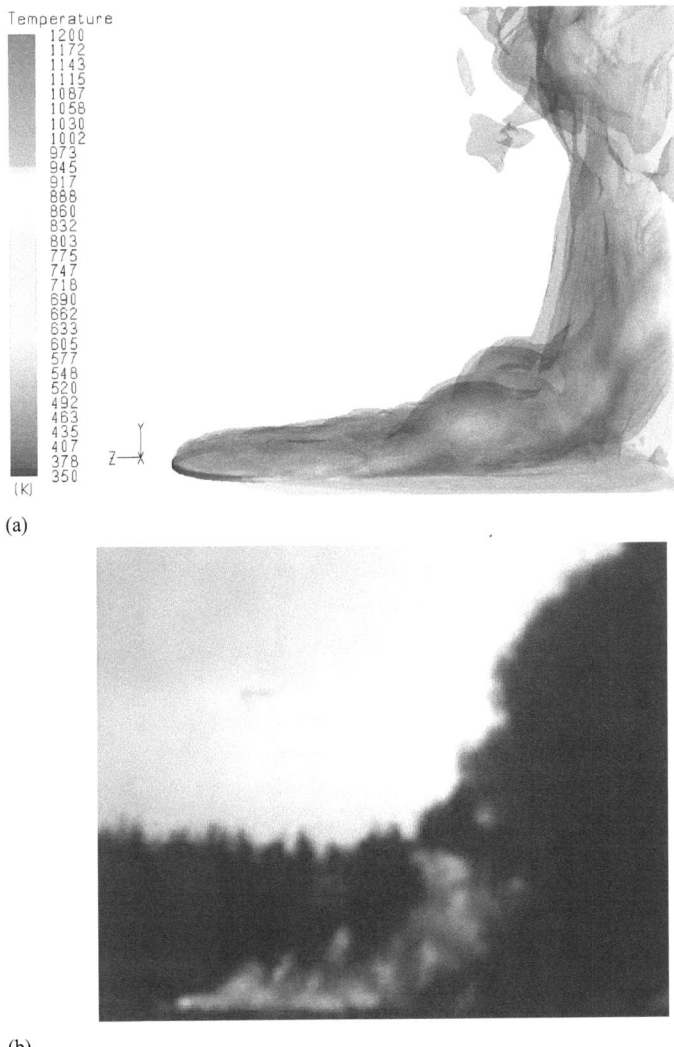

Fig. 5.20: (a) CFD predicted isosurfaces of temperatures (400 K < T < 1400 K) and (b) VIS image of a JP-4 pool fire (d = 2 m) under the influence of the cross-wind (u_w = 4.5 m/s).

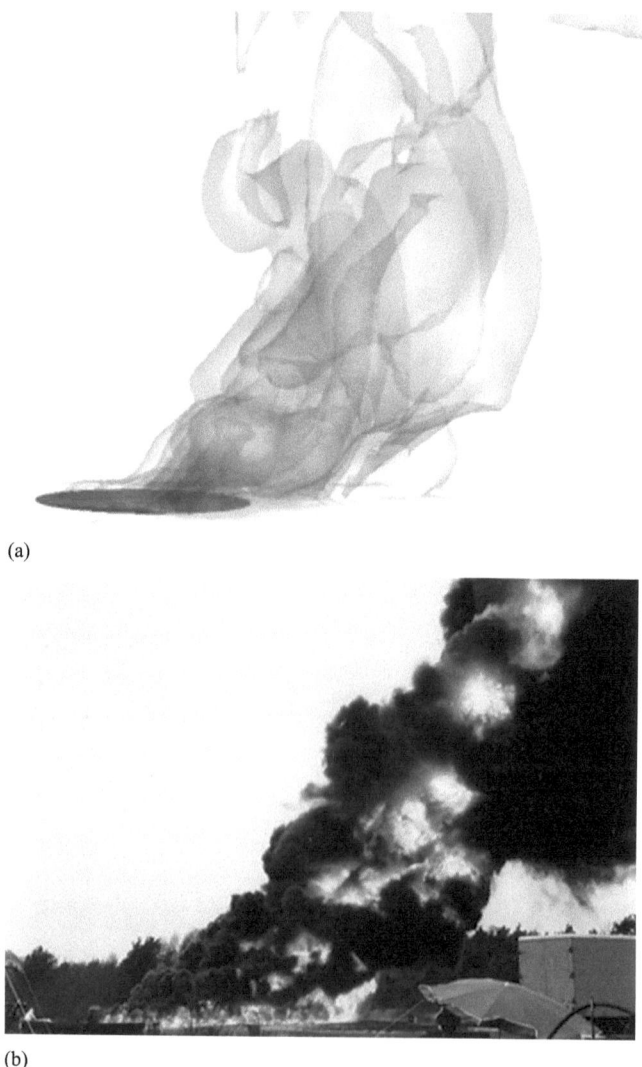

(a)

(b)

Fig. 5.21: (a) CFD predicted isosurfaces of temperature (400 K < T < 1400 K) and (b) VIS image of a JP-4 pool fire (d = 25 m) under the influence of the cross-wind (u_w = 4.5 m/s).

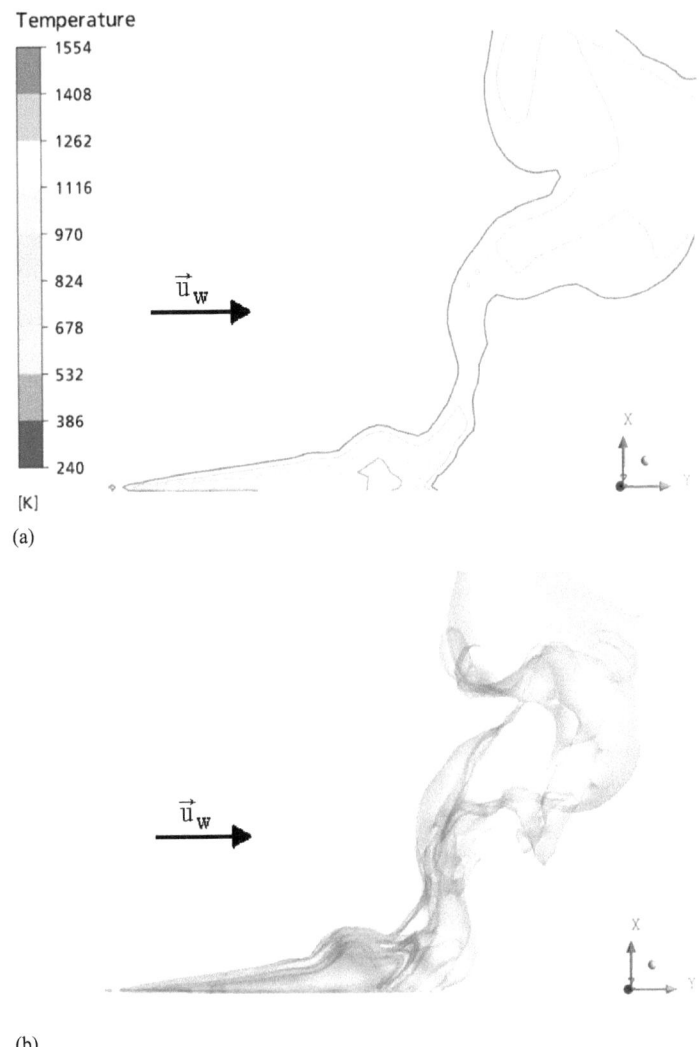

Fig. 5.22: CFD predicted (a) isotherms and (b) isosurfaces of temperatures (400 K ≤ T ≤ 1000 K) of JP-4 pool fire (d = 20 m) under the influence of the cross wind (u_w = 10 m/s).

168   5 Results and discussions

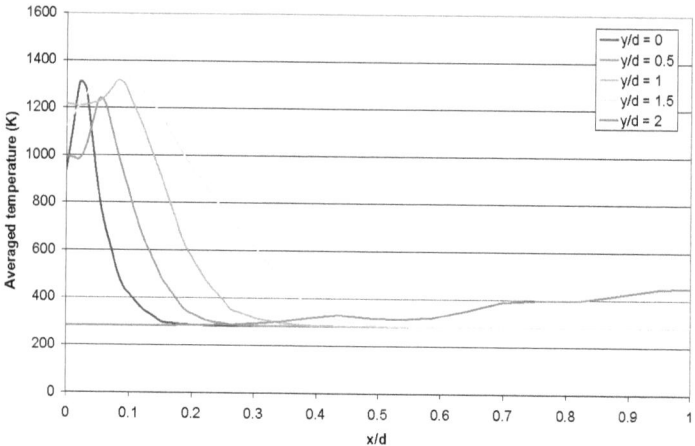

Fig. 5.23: CFD predicted time averaged axial temperature profiles at different radial distances y/d from the flame axis of JP-4 pool fire (d = 20 m) under the cross-wind (u_w = 10 m/s).

Wind influence on flow velocity

In the CFD simulations it is not found a significant increasing of the flame velocity but it is predicted that the maximum flow velocity in a flame always follows the flame tilt and is found in a plume region (Fig. 5.24, 5.25).

The additional wind flow in a simulation influences the rotational velocities so the vortex formation is more pronounced than in simulations without wind included even if the RANS calculation and simplified chemistry (multistep chemical reaction) is used.

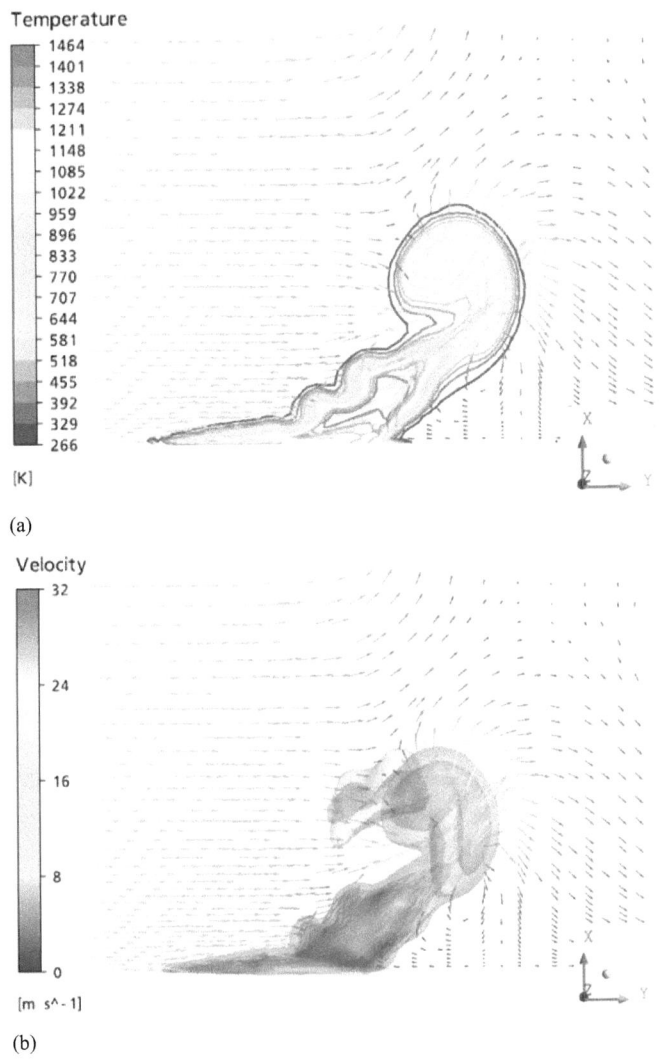

Fig. 5.24: CFD predicted (a) instantaneous isotherms and flow field and (b) isosurfaces of temperatures (400 K ≤ T ≤ 1000 K) and flow field of JP-4 pool fire (d = 20 m) under the influence of the cross wind of u_w = 7.2 m/s at t = 8 s.

170   5 Results and discussions

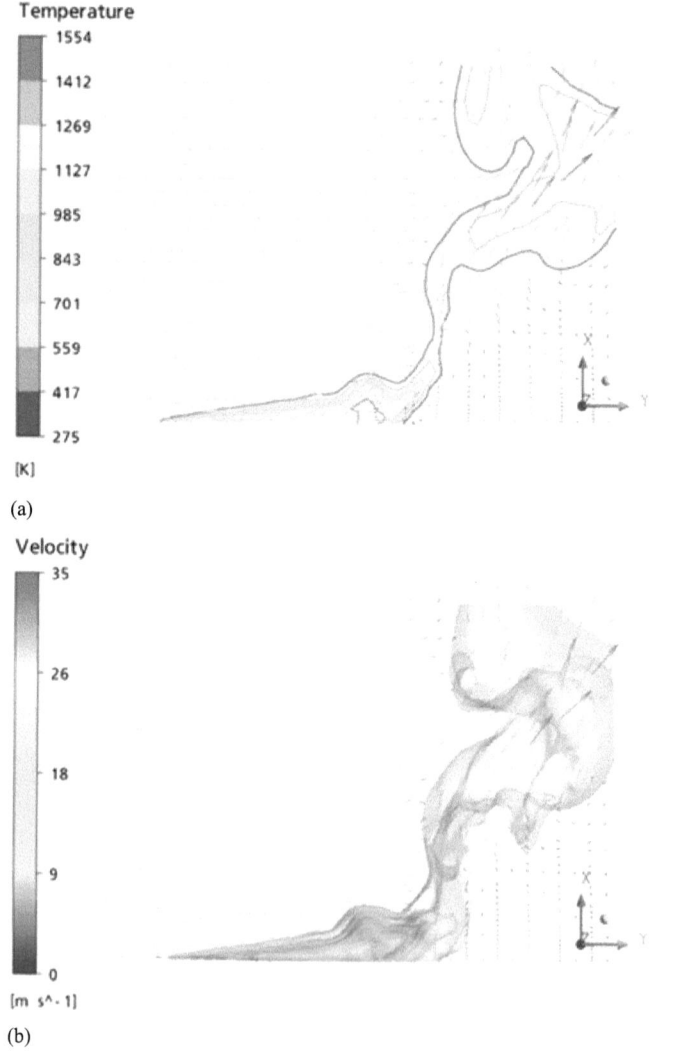

(a)

(b)

Fig. 5.25: CFD predicted (a) instantaneous isotherms and flow field and (b) isosurfaces of temperatures (400 K ≤ T ≤ 1000 K) and flow field of JP-4 pool fire (d = 20 m) under the influence of the cross wind of 16 m/s at t = 8 s.

5.5 Validation of CFD results

The validation of the calculated fields and sizes, such as temperature, flow velocities, thermal radiation and soot amount in large JP-4 pool flames contains several steps. First is to use the images and profiles of the transient values to give the relationship between the individual fields sizes (Chapter 5.1). Next step is to compare the predicted time averaged temperatures with the measured data. The following is to look at the transient thermal radiation such as incident radiation to the chosen area or irradiances at the different distances from the pool rim (Chapter 5.2 and 5.3) to discuss the influence of the flame dynamic on the thermal radiation from the fire. Further step is to look into transient data of soot amount in large fires to explain the influence on the thermal radiation. In the next step the CFD predicted time averaged values of incident radiations and irradiances are compared with measurements or empirical correlations to discuss about the strengths of the flame radiation and its effect on neighboring objects. The simple juxtaposition of CFD simulations and experiments can be strictly taken as a validation. Due to the lack of clear definitions for validation and verification, this method is however a long time been considered sufficient and is still very widespread. More recently, the desire is to identify more accurate estimates for the experimental and numerical uncertainties. According to the American Institute of Aeronautics and Astronautics (AIAA) is in the verification examined whether an implemented model is suitable for the certain concept and if the implemented model correctly solves its equations [108]. During the validation is examined how accurately a model represents the part of the "real world" for which it was developed.

For the quantitative evaluation of simulations in [108] different criteria are defined to indicate the errors in relation to a size x. The error $F(x)$ depends directly on the calculated value $y_m(x)$ and the mean value $y_e(x)$ from the experiments from:

$$F(x) = y_m(x) - y_e(x). \tag{5.16a}$$

The error is with a 90% probability in the field

$$\left(F(x) - t_{0.05,f} \frac{s(x)}{\sqrt{n}}, F(x) + t_{0.05,f} \frac{s(x)}{\sqrt{n}} \right) \tag{5.16b}$$

where $s(x)$ is standard deviation. The distribution of t depends on the degrees of freedom $f = n - 1$ for n of x.

Oberkampf et.al gives also the criteria for the average and maximum error:

$$\left|\frac{F}{\overline{y}_e}\right|_{avg} = \frac{1}{x_u - x_1} \int_{x_1}^{x_u} \left|\frac{y_m(x) - \overline{y}_e(x)}{\overline{y}_e(x)}\right| dx \qquad (5.16c)$$

$$\left|\frac{F}{\overline{y}_e}\right|_{max} = \max_{x_1 \le x \le x_u} \left|\frac{y_m(x) - \overline{y}_e(x)}{\overline{y}_e(x)}\right| \qquad (5.16d)$$

The highest value of size x is here with x_u and with the lowest x_1 are described. An average relative confidence indicator (CI) is calculated by:

$$\left|\frac{CI}{\overline{y}_e}\right|_{avg} = \frac{t_{0.05v}}{(x_u - x_1)\sqrt{n}} \int_{x_1}^{x_u} \left|\frac{s(x)}{\overline{y}_e(x)}\right| dx \qquad (5.16e)$$

The maximum CI is calculated by:

$$\left|\frac{CI}{\overline{y}_e}\right|_{max} = \frac{t_{0.05v}}{\sqrt{n}} \int_{x_1}^{x_u} \left|\frac{s(\hat{x})}{\overline{y}_e(\hat{x})}\right| dx \qquad (5.16f)$$

Using these equations the deviations of the simulation results from the experiment can be quantitatively assessed. The knowledge of the measurements and their associated uncertainties is necessary. It should also measurements of various working groups and correlations be used but their uncertainties, however, generally are not known. In this work a directly comparison of the CFD results with the measured data is presented in a form of fields, profiles and time average values due to the reason that at this time are still too little experimental data available.

In the following text validation of CFD results of temperature and SEP of JP-4 pool fires is done by using global metric [108].

Calculation of uncertainties of the flame temperature and SEP

By using global metric (Eqs. (5.16a)-(5.16f)) the error is calculated for CFD predicted $\overline{T}$ and $\overline{SEP}$ of JP-4 pool fire (d = 16 m) compared with experimental values from Tab. 5.1 and Fig. 5.14. The results are shown in Tab. 5.10a,b.

CFD results in Tab. 5.10a,b show a good agreement with the measured values. Averaged and maximal relative errors are small due to the large confidence interval CI. Code validation needs more experimental data to decrease CI.

Table 5.10a: Calculation of uncertainties of the time averaged temperature $\overline{T}$ of JP-4 pool fire (d = 16 m)

	SAS (1M)*
Averaged relative error	0.14
Averaged relative confidence interval	0.89
Maximal relative error	0.79
Confidence interval of max. error	0.89

*CFD simulation is done by using SAS model and 1 million cell mesh

Table 5.10b: Calculation of uncertainties of the time averaged $\overline{SEP}$ of JP-4 pool fire (d = 16 m)

	SAS (1M)*
Averaged relative error	0.118
Averaged relative confidence interval	0.89
Maximal relative error	0.95
Confidence interval of max. error	0.89

*CFD simulation is done by using SAS model and 1 million cell mesh

6. Conclusions

The following conclusions are summarized from the results:
1. CFD simulation on sooty, large, hydrocarbons pool fires develops to a powerful method for a prediction of thermal radiation from the fire.
2. By CFD simulation it is possible to predict the SEP_{CFD} as a function of time and space. The "derived" quantity SEP_{CFD} can be predicted by using following three ways:
   - Determination of an isosurface of constant temperature as a flame surface
   - Integration over many distributions of incident radiation G along the z direction through the flame
   - By irradiance $E(\Delta y/d, t)$ calculated by virtual wide angle radiometers defined at the pool rim $\Delta y/d = 0$.
3. CFD simulation predicts also the time dependent irradiances $E_{CFD}(\Delta y/d, t)$ and time averaged $\overline{E}_{CFD}(\Delta y/d)$ by virtual radiometers at different horizontal distances $\Delta y/d$ from the pool rim.
4. A four-step discontinuity function for the effective absorption coefficient $\overline{\overline{æ}}_{eff}$ which considers the dissipative structures reaction zones, hot spots and soot parcels is presented.
5. By CFD simulation it is possible to predict the flame tilt, drag, temperature, SEP and irradiance of the pool fires under the wind influence.
6. For the successful CFD simulation it is necessary to use a detailed reaction mechanism as flamelet models.
7. Further progress is focused on a reduction of CPU time e.g. by using of compute cluster with higher performance.

References

[1] M. Hailwood, M. Gawlowski, B. Schalau, A. Schönbucher, *Chem. Eng. Technol.* (2009), doi:10.1002/ceat.200800595.

[2] *The Buncefield Investigation, Progress Report*, February 2006, www.buncefieldinvestigation.gov.uk/report.pdf.

[3] D. Göck, Experimentell fundierte Ballenstrahlungsmodelle zur Bestimmung von Sicherheitsabständen bei großen Poolflammen flüssiger Kohlenwasserstoffe, *Dissertation*, Universität Stuttgart, 1988.

[4] R. Fiala, D. Göck, X. Zhang, A. Schönbucher, *Chem. Ing. Tech.* 63 (1991) 760 – 761.

[5] K.S. Mudan, *Prog. Energy Comb. Sci.*, 10 (1984) 59 – 80.

[6] F. P. Lees, *Lee's Loss Prevention in the Process Industries* (Eds. S. Mannan), 3th ed., Elsevier Butterworth-Heinemann, Burlington, MA, 2005.

[7] A. Schönbucher, *Quellterme bei offenen Bränden von Flüssigkeiten und Gasen*, interner Bericht 2008.

[8] H. Koseki, Y. Iwata, *Fire Technol.*, 36, 1 (2000) 24 – 38.

[9] S.R. Tieszen, V.F. Nicolette, L.A. Gritzo, J.K. Holen, D. Murray, J.L. Moya, Vortical structures in pool fires: observation, speculation, and simulation, Sandia report, SAND96-2607, November 1996.

[10] M. Gibbons, C. Devaud, E.J. Weckman, Behaviour of Pool Fires in a Crosswind: Comparison of Experimental and Computational Results, *CS/The Combustion Institute*, Halifax, May, 6 (2005).

[11] T. T. Fu, *Fire technology*, 10-1 (1974) 54 – 67.

[12] S.R. Tieszen, *Annual Review of Fluid Mechanics* (2001) 33 – 67.

[13] H. Koseki, *Proc. of the 2nd Int. Symp. on Scale Modeling*, Lexington 1997.

[14] K.B. McGrattan, H.R. Baum, A. Hamins, Thermal Radiation from Large Pool Fires, *Report NISTIR* 6546, Fire Saftey Engineering Division Building and Fire research Laboratory, 2000.

[15] K.B. McGrattan, H.R. Baum, R.G. Rehm, *Fire Safety Journal* (1998) 30 –161.

[16] M. Faghri, B. Sundén, Transport phenomena in fires, WIT press, Southampton, Boston, 2008, ISBN: 978-1-84564-160-3.

[17] Y.L. Sinai, *Fire Safety Journal* 35 (2000) 51 – 61.
[18] V.F. Nicolette, L.A. Gritzo, *Proc. of the 4th Int. Symp. on Fire Safety Science*, Ottawa (1994).
[19] Y.L. Sinai, *Fire Safety Journal*, 24, 1 (1995) 1 – 34.
[20] M. Muñoz, J. Arnaldos, J. Casal, E. Planas, *Comb. Flame*, 139 (2004) 263 – 277.
[21] M. Muñoz, E. Planas, F. Ferrero, J. Casal, *J. Hazard. Mater.* 144 (2007) 725 – 729.
[22] J.A. Fay, *J. Hazard. Mater.* B 136 (2006) 219 – 232.
[23] P.K. Raj, *J. Hazard. Mater.* 140 (2007) 280 – 292.
[24] P.K. Raj, *Process Safety Progress* 24, 3 (2005) 192 – 202.
[25] W.F.J.M. Engelhard in *Methods for the Calculation of Physical Effects – The Yellow Book* (C.J.H. van den Bosch, R.A.P.M. Weterings (Eds.)), 3rd Ed., Part 2, 6.1-6.130, The Committee for the Prevention of Disasters, TNO, The Hague, The Netherlands, 1997.
[26] M. Gawlowski, M. Hailwood, I. Vela, A. Schönbucher, *Chem. Eng. Technol.* (2009), doi:10.1002/ceat.200800631.
[27] I. Vela, H. Chun K. B. Mishra, M.Gawlowski, P. Sudhoff, M. Rudolph, K.-D. Wehrstedt, A. Schönbucher, *Forschung Ingenieurwes.*, 2009, in press.
[28] C. Kuhr, CFD-Simulation der dynamischen Eigenschaften großer Kerosin- und Heptan-Poolflammen, *Dissertation*, Universität Duisburg-Essen, 2008.
[29] K.A. Jensen, J.-F. Ripoll, A.A. Wray, D. Joseph, M. El Hafi, *Combustion and Flame* 148 (2007) 263 – 279.
[30] C. Balluff, *VIS-Ballenstrukturen und Oszillationen in Großflammen*, Dissertation, Universität Stuttgart, 1989.
[31] H. Chun, *Experimentelle Untersuchungen und CFD-Simulation von DTBP-Poolfeuern*, BAM-Dissertationsreihe, Band 23, Berlin, 2007, www.bam.de/de/service/publikationen/publikationen_medien/diss_23_vt.pdf
[32] C.H. Hottel: Certain Laws Covering Diffusive Burning of Liquids, *F. Res. Abs. And Rev., Vol.* 3 (1959), 41 – 44.
[33] H.G. Werthenbach, *Z. Forschung Technik Brandschutz* 25 (1976) 57
[34] M. Hertzberg, *Combustion and flame*, 21 (1973) 195 – 209.

[35] A. Schönbucher, Massenabbrandraten und Lachengrößen bei Flüssigkeitsbränden, presentation at 648. *DECHEMA colloquium*: "Entwicklung von Quelltermen für Auswirkungsbetrachtungen", Dechema Haus, Frankfurt am Main, February 12. 2009.

[36] D. Burgess; A. Strasser; J. Grumer, *Fire Res. Abst. Revs.* 3 (1961) 177 – 192.

[37] J. Grumer, A. Strasser, T.A. Kubala, D.S. Burgess, Fire Res. Abstr. Rev. 3 (1961) 159

[38] D.S. Burgess, J.N. Murphy, M.G. Zabetakis, Bureau of mines Pittsburg PA Safety Research Center, Repository Defense Technical Information Center OAI-PMH Repository (United States) (2005)

[39] V. Babrauskas, Book Review. Fire Investigation, *Fire Safety J.* 40 (2005) 299 – 300

[40] V.I. Blinov and G.N. Khudiakov, Certain Laws Governing Diffusive Burning of Liquids, Academiia Nauk, SSSR Doklady, Vol. 113, 1957, pp. 1094 – 1098.

[41] L. Orloff: Simplified Radiation Modeling of Pool Fires, Proc. Combust. Inst., Vol. 18 (1972) 549 – 583

[42] P.H. Thomas, The size of flames from natural fires, in: Proceedings of the 9th International symposium on Combustion, The Combustion Institute, 1963, pp. 844 – 859.

[43] F.R. Stewart, *Comb. Sci. Technol.*, Vol. 2 (1970) 203 – 212.

[44] J. Moorhouse, *The Assessment of Major Hazards*, Institution of Chemical Engineers Symposium Series No. 71, Rugby (1982) 165 –180.

[45] G. Heskestad, *Fire Safety J.* 5 (1983) 103 –108

[46] E.E. Zukoski, B.M. Cetegen, T. Kubota, *The Combustion Institute* (1985) 361 – 366.

[47] B.J. McCaffrey, Purely Buoyant Diffusion Flames: Some Experimental Results, *Final Report*, NBSIR 79-1910, 49 p, October 1979,

[48] J.R. Welker, C.M. Sliepcevich, *Wind interaction Effects on Free Burning Fires*, Tech. Report #1441-3 to Office of Civil Defense of U.S. Bureau of Standards, 1967

[49] American Gas Association, *LNG Safety Research Programm*, Report IS 3-1, 1974

[50] E.J. Weckmann, A.B Strong, *Comb. Flame* 105 (1996) 245 – 266

[51] B.M. Cetegen, *Combust. Sci. Tech.* 123 (1997) 377 – 387

[52] S. Venkatesh, A. Ito, K. Saito, I.S. Wichman, *26th Symp. (Int.) Comb.* (1996), The Combustion Institute, Pittsburgh, 1437 – 1443

[53] Y. Xin, J.P. Gore, K.B. McGrattan, R.G. Rehm, H.R. Baum, *Comb. Flame* 141 (2005) 329 – 335

[54] C. Kuhr, D. Opitz, R. Müller, A. Schönbucher, *Proc. 10th Int. Symp. On Loss Prevention and Safety Promotion in the Process Industries* (2001), June 19-21, Stockholm, Sweden, 1179 – 1188

[55] C. Kuhr, D. Opitz, R. Müller, A. Schönbucher, *5. Fachtagung Anlagen-, Arbeitsund Umweltsicherheit* (2000), Köthen, GVC.VDI-Gesellschaft Verfahrenstechnik und Chemieingenieurwesen, 311 – 318

[56] C. Kuhr, *Transiente Felder der Strömungsgeschwindigkeit, Vortizität und Zirkulation in großen JP4-Poolflammen*, Diplomarbeit, Universität Duisburg, 2000

[57] H. Koseki, T. Yumoto, *Fire Technol.* 24 (1988) 33-47

[58] E. Planas-Cuchi, J. Casal, *J. Hazard. Mat.* 62 (1998) 231 – 241.

[59] B.L. Bainbridge, N.R. Keltner, *J. Hazar. Mater.* 20 (1988) 21 – 40.

[60] J.J. Gregory, N.R. Keltner, R.J. Mata, *J. Heat Transfer-Transactions of the ASME* 111 (1989) 446 – 454.

[61] K.B. McGrattan, H.R.Baum, A. Hamins, *Thermal Radiation from Large Pool Fires*, Report NISTIR 6546, Fire Saftey Engineering Division Building and Fire research

[62] M. Frenklach, D.W. Clary, W.C. Gardiner, S.E. Stein, *21st Symp. (Int.) Comb.* (1986), The Combustion Institute, Pittsburgh, 632 – 647.

[63] H. Bockhorn, T. Schäfer, in *Soot Formation in Combustion* (Bockhorn, H. (Ed.)), 253 – 274, Springer Berlin, 1994.

[64] R. Dobbins, R.A. Fletcher, H.C. Chang, *Comb. Flame* 115 (1998) 285 – 298.

[65] H. Bockhorn, in *Soot Formation in Combustion* (Bockhorn, H. (Ed.)), 3–7, Springer Berlin, 1994.

[66] U. Vandsburger, I. Kennedy, I. Glassman, *Combust. Sci. Technol.* 39 (1984) 263 – 285.

[67] A. Levy, *19th Symp. (Int.) Comb.* (1982), The Combustion Institute, Pittsburgh, 1223 – 1242.

[68] J. Warnatz, U. Maas, R.W. Dibble, *Verbrennung*, 4. Auflage, Springer, Berlin Heidelberg, 2006

[69] D. Bradley, G. Dixon-Lewis, S.E.D. Habik, E.M.J. Mushi, *20th Symp. (Int.) Comb.* (1984), The Combustion Institute, Pittsburgh, 931 – 940

[70] C.P. Fenimore, G.W. Jones, *J. Phys. Chem.* 71 (1967) 593 – 597

[71] F.P. Lees, in: S. Mannan (Ed.), Loss Prevention in the Process Industries, vol. 2, 3rd ed., Elsevier Butterworth-Heinemann, Burlington, MA, 2005.

[72] P.J. Rew, W.G. Hulbert, Development of Pool-Fire Thermal Radiation Model, *HSE Contract Research Report* No. 96 / 1996, ISBN 0717610845

[73] W.K. Moser and C.F. Moser In: Fire and Forest Ecology: Innovative Silviculture and Vegetation Management, (Eds.). Proceedings 21st Tall Timbers Fire Ecology Conference on Fire and Forest Ecology, Tallahassee, Florida, 1998.

[74] L.T. Cowley, A.D. Johnson (1991), in [25]

[75] C.L. Beyler, in: P.J. DiNenno et al. (Eds.), SFPE Handbook of Fire Protection Engineering. National Fire Protection Association, Quincy, MA, Section 3, chap. 11, 3-268 to 3-314, 3 (2002).

[76] S. Mannan, Lee's Loss Prevention in the Process Industries, Vol. 1-3, Elsevier Butterworth-Heinemann, 3 (2005).

[77] P.G. Seeger, GWF, Gas Erdgas 120 (1979) 25.

[78] A.N. Kolmogorov, *J. Fluid Mech.* 13 (1962) 82 – 85

[79] M. Griebel, T. Dornseifer; T. Neunhoeffer, *Numerische Simulation in der Strömungsmechanik*, Vieweg, Braunschweig, 1995

[80] E.S. Oran, J.P. Boris, *Numerical Simulation of Reactive Flow*, 2nd Ed., Cambridge University Press, 2001

[81] R. Bryant, C. Womeldorf, E. Johnsson, T. Ohlemiller, *Fire Mater.* 27 (2003) 209 – 222

[82] I.G. Currie, *Fundamental Mechanics of Fluids*, 2nd Ed., McGraw-Hill, 1993

[83] ANSYS CFX 11.0 User Guide. AEA Technology, 2008, www.ansys.com

[84] ANSYS FLUENT 12 User Guide. AEA Technology, 2008, www.ansys.com

[85] A. Schönbucher: Thermische Verfahrenstechnik Grundlagen und Berechnungsmethoden für Ausrüstungen und Prozesse, Springer-Verlag Berlin Heidelberg, ISBN 3-540-42005-3 (2002)

[86] P.A. Libby, F.A. Williams, *Turbulent Reacting Flows*, Academic Press, London, 1994

[87] J.H. Ferziger, M. Peric, *Computational Methods for Fluid Dynamics*, 2nd Edit., Springer, Berlin Heidelberg, 1997

[88] A.R. Paschedag, *CFD in der Verfahrenstechnik*, WILEY-VCH Weinheim, 2004

[89] Y. Egorov and F. Menter, Development and Application of SST-SAS Turbulence, Advances in Hybrid RANS-LES Modelling in: *Notes on Numerical Fluid Mechanics and Multidisciplinary Design*, Springer Berlin/Heidelberg, Volume 97/2008, ISBN: 1612-2 909 (Print) 1060-0824 (Online) DOI: 10.1007/978-3-540-77815-8

[90] E. Riesmeier, S. Honnet, N. Peters, *Flamelet Modeling of Pollutant Formation in a Gas Turbine Combustion Chamber using Detailed Chemistry for a Kerosene Model Fuel*, Proceedings of the 2002 Fall Technical Conference of the ASME Internal Combustion Engine Division, ICE-Vol. 39 (2002) 149 – 157.

[91] N. Peters, G. Paczko, R. Seiser, K. Seshadri, *Comb. Flame* 128 (2002) 38 – 59.

[92] B.F. Magnussen, *Modelling of NOx and Soot Formation by the Eddy Dissipation Concept*, Presented at the First Topic Oriented Technical Meeting, International Flame Research Foundation, Amsterdam, The Netherlands, 1989

[93] P.A. Tesner, T.D. Snegirova, V. G. Knorre, *Comb. Flame* 17 (1971) 253 – 260

[94] K.M. Leung, R.P. Lindstedt, Jones, W.P., *Comb. Flame* 87 (1991) 289 – 305

[95] M. Fairweather, W.P. Jones, R.P. Lindstedt, *Comb. Flame* 89 (1992) 45 – 63

[96] J.H. Miller, W.G. Mallard, K.C. Smyth, *22st Symp. (Int.) Comb.* (1986), The Combustion Institute, Pittsburgh, 1057 – 1065

[97] J.H. Kent, H.G. Wagner, *Comb. Flame* 47 (1982) 53 – 65

[98] U. Vandsburger, I. Kennedy, I. Glassman, Combust. Sci. Technol. 39 (1984) 263 – 285

[99] F. Liu, H. Guo, G.J. Smallwood, Ö.L. Gülder, *Combust. Theory Modelling* 7 (2003) 301 – 315

[100] K.B. Lee, M.W. Thring, J.M. Beer, *Comb. Flame* 6 (1962) 137 – 145

[101] M. Griebel, T. Dornseifer, T. Neunhoeffer, *Numerische Simulation in der Strömungsmechanik*, Vieweg, Braunschweig, 1995

[102] J.M. Chatris, J. Quintela, J. Folch, E. Planas, J. Arnaldos, J. Casal, *Comb. Flame* 126 (2001) 1373 – 1383

[103] H. Koseki, G.W. Mulholland, T. Jin, Study on Combustion Property of Crude Oil - A Joint Study Between NIST/CFR and FRI, *Eleventh Joint Panel Meeting of the UJNR Panel on Fire Research and Safety*. Online im Internet: http://www.mms.gov/tarprojects/026.htm, 2006

[104] J. Fröhlich: Large Eddy Simulation turbulenter Strömungen, Teubner-Verlag, ISBN 10 3-8351-0104-8 (2006)

[105] K.M. Leung, R.P. Lindstedt, W.P. Jones, *Comb. Flame* 87 (1991) 289–305

[106] C.L. Tien, K.Y. Lee, A.J. Stretton, Radiation heat transfer, in: *SPFE handbook of fire protection engineering*, 2nd ed. Quincy, MA: NationalFire Protection Association, 1995.

[107] C.W. Lautenberger, J.L. de Ris, N.A. Dembsey, J.R. Barnett, H.R. Baum, *Fire Safety J.* 40 (2005) 141 – 176.

[108] W.L. Oberkampf, M.F. Barone, *J. Comp. Phys.* 217 (2006) 5 – 36.

References

Die VDM Verlagsservicegesellschaft sucht für wissenschaftliche Verlage abgeschlossene und herausragende

Dissertationen, Habilitationen, Diplomarbeiten, Master Theses, Magisterarbeiten usw.

für die kostenlose Publikation als Fachbuch.

Sie verfügen über eine Arbeit, die hohen inhaltlichen und formalen Ansprüchen genügt, und haben Interesse an einer honorarvergüteten Publikation?

Dann senden Sie bitte erste Informationen über sich und Ihre Arbeit per Email an *info@vdm-vsg.de*.

Sie erhalten kurzfristig unser Feedback!

VDM Verlagsservicegesellschaft mbH
Dudweiler Landstr. 99
D - 66123 Saarbrücken
www.vdm-vsg.de

Telefon +49 681 3720 174
Fax +49 681 3720 1749

Die VDM Verlagsservicegesellschaft mbH vertritt

Printed by Books on Demand GmbH, Norderstedt / Germany